# A CENTURY
# OF FLUID MECHANICS
# IN THE NETHERLANDS

# A CENTURY
# OF FLUID MECHANICS
# IN THE NETHERLANDS

## FONS ALKEMADE

 Springer

Fons Alkemade
Haarlem, The Netherlands

ISBN 978-3-030-03585-3      ISBN 978-3-030-03586-0   (eBook)
https://doi.org/10.1007/978-3-030-03586-0

Manuscript Design and layout: Curve Mags and More, Haarlem, Holland, www.curve.nl.
Cover design: Curve Mags and More Photo: Hollandse Hoogte/Siebe Swart

This Springer imprint is published by the registered company Springer Nature Switzerland AG.
The registered company address is: Gewerbestrasse 11, 6330 Cham, Switzerland

# PREFACE

The publication of this book marks the centenary of the appointment of the first professor in fluid mechanics in the Netherlands: this was Johannes (Jan) Martinus Burgers (1895–1981), who was appointed at the age of 23 (!) as professor at Delft University of Technology. This historical fact forms a good reason to look back on 100 years of fluid mechanics in the Netherlands. In particular, on developments in research in this area, both in academic settings and in industrial laboratories.

Fluid mechanics as an engineering activity has been practised in this country for much longer – just think of the development of wind mills for various purposes, the building of waterways and dikes for water management and coastal protection, land reclamation, ship building, and other engineering applications. Its broad societal relevance and its many direct applications have always been a primary driving force in the development of fluid mechanics as a branch of engineering. A century ago, it was realised that a more fundamental study of this field was also relevant and much needed, in order to lay a firm foundation under this fast-developing field. Research on fundamental issues would become of great importance for essential, deeper insights into the dynamics of the various flow phenomena encountered in daily life, be they in natural situations or in industrial configurations.

In this book, Fons Alkemade has collected interesting historical material that illustrates the activities of the various branches of fluid dynamics, thus providing a comprehensive overview of the development of the field since the appointment of J.M. Burgers. Many new areas have emerged within the field, and have led to essential, better understanding of the fundamental dynamics. Such important new developments were made possible by remarkable progress in experimental techniques and increasingly powerful computational resources.

This book project has received financial support from a number of industrial sponsors, and also from the JM Burgerscentrum (JMBC), the national research school for fluid mechanics in the Netherlands. This school is proud to be named after the first chair holder in fluid mechanics in this country. The JMBC acts as an umbrella organisation for many academic fluid dynamics research groups in different university departments, ranging from chemical, civil, and mechanical engineering to mathematics, physics, and zoology – illustrating the wide relevance of the field. With about 200 scientific staff and more than 350 PhD students and post-docs participating, it nicely shows that in the Netherlands fluid mechanics is a lively and very active branch of science and engineering.

It is my wish that reading this book will be as great a pleasure for you, as it has been for me.

*GertJan van Heijst*
*scientific director JM Burgerscentrum*

# ACKNOWLEDGMENTS

I want to thank the following persons for:

- the talks and interviews I had with them: Jurjen Battjes, Wim-Paul Breugem, René Delfos, Bram Elsenaar, Martien Hulsen, Geert Keetels, Gijs Ooms, Christian Poelma, Mathieu Pourquie, Cornel Thill, Wim Uijttewaal, Willem van de Water, Jerry Westerweel

- sharing with me all the names, dates, facts from his amazing memory: Leen van Wijngaarden, whose career, together with that of his father C.M. van Wijngaarden, professor in dredging engineering in Delft from 1922 onwards, covers almost the complete period treated in this volume

- reading parts of the manuscript: Joost den Haan, Ruud Henkes, Jos Zeegers

- improving my English: Leo Finn

- helping me with illustrations and/or giving me additional information: Lourens Aanen (Peutz), Gerard Alberts (UvA), Hans van Bergem (Deltares), René Bevers (UT), Jaap Boelens (RWS), Onno Bokhove (University of Leeds), Daniel Bonn (UvA), Flip Bool, Arno Brand (ECN), Leo Broers (HHNK), Bert Brouwers (Romico), Eelco Brouwers (Hist. Centrum Overijssel), Billa de Bruijn (DSM), Gustave Corten (CortEnergy), Remmelt Daalder, Marion Dietz, Anton Dommerholt, Cas van Doorne, Bram Elsenaar, Geertje van Emmerik (De Cruquius), Erik Geelen (TUE), Frans van Grunsven (TUD), Rob Hagmeijer (UT), Willem van der Ham, Rikkert Harink, Martin van Hecke (UL), Carlo van der Heijden, Ruud Henkes (TUD / Shell), Aad Hermans, Mico Hirschberg, Ilse Hoekstein-Philips (Burgerscentrum), Wim van Hoeve (Tide Microfluidics), Ton Hoitink (WUR), Giel van Hooff, Sander Huisman (UT), Jan van Ingen, Alex Jansen (Interflow), Jos Jansen, Tom de Kievith (UU), Maarten Kleinhans (UU), Wouter Knap (KNMI), Carry Koolbergen (IOS Press), Gerard Kuiken, Rudie Kunnen (TUE), Leo Maas (UU), Roy Mayer (FlowMotion), Charlotte van Mierlo (Océ), Bennie Mols, Rob Mudde (TUD), Hans van Muiswinkel, Kiri Nichol, Dries van Nimwegen, Gijs van Ouwerkerk, Bart van Overbeekde, Gerrit Peters (TUE), Peter Poot (TUD), Thijs Postma, Ed van Rijswijk, Leandro De Santana (UT), Roland Schmehl (TUD / Kitepower), Jan Smeulers (TNO), Trienke van der Spek (Teylers Museum, Haarlem), Alexander Spoelstra (TUD), Adrian Steketee, Marcel Stive (TUD), Lucy Straker (IFRF), Simone Straten (Peutz), Abel Streefland (TUD), Jaap den Toonder (TUE), Arthur Veldman (RUG), Kees Venner (UT), Michel Versluis (UT), Frank Visser (Flowserve), Jeroen van der Vliet (Scheepvaartmuseum Amsterdam), Jack Voncken, Bert Vreman (Akzo Nobel / TUE), Co de Vries, Marco de Waal (De Waal bv), Gerard Wegdam (UvA), Jebbe van der Werf (Deltares), Ellen te Winkel (MARIN), Paul van Woerkom (TUD), Yonghui Zhu

Special thanks to Robert Byron Bird ('Bob de Vogel', 1924) for contributing to this book.

Thanks to the Stichting Pieter Zeeman-Fonds (Anne Kox) for their financial support and trust in me. Thanks to the sponsors from industry and research. They have not only made possible this book financially but have also contributed some pages on their activities related to fluid mechanics: MARIN (Henk Prins), Océ (Hans Reinten), Shell (Peter Veenstra and Ruud Henkes), Tata Steel (Tim Peeters), Teijin Aramid (Hans Meerman), Unilever (Jo Janssen), VSL (Joost Groen). The author of this volume cannot be held accountable for the content of these pages.

Thanks to GertJan van Heijst of the J.M. Burgerscentrum. Without his positive reaction to my proposal and his continuing interest, comments and encouragement, this book would never have been made.

And last but not least, many thanks to Henk Stoffels and Roy Wolfs of Curve (www.curve.nl) who have done an excellent job in giving this volume its lay-out and preparing it for publication.

# CONTENTS

# 01

J.J. de Vries

## INTRODUCTORY REMARKS

Try to imagine: you have just been appointed as a professor in in a field of science of which you know little and of which you are the very first in your country. You have no laboratory, you have no international contacts and you are only 23 years old.

We do not know what Jan Burgers thought of his situation in his first years as professor in Delft, but we do know that he took his job very seriously and soon became well-known and respected in the world of mechanics. In 1975, contemplating on his years in Delft (1918–1955) in the *Annual Review of Fluid Mechanics*, he commented: "Looking back, I may even say that the major part of my scientific work has been directed towards interpretation, more than to finding new results, although interpretation often opens the mind for a new view."

8

© Springer Nature Switzerland AG 2019
F. Alkemade, *A Century of Fluid Mechanics in The Netherlands*, https://doi.org/10.1007/978-3-030-03586-0_1

↑One of the few places in the Netherlands where the wonderful world of fluid flows is regularly presented to 'laymen' (and where they are invited to think about this world) is the column 'Alledaagse Wetenschap' (Everyday Science) written by science editor Karel Knip in the newspaper *NRC Handelsblad*. Knip started his weekly series in 1991 and the number of issues in which fluid mechanics is involved must by now have passed one hundred. Topics range from the waves generated by ducks or raindrops, to the rising smoke of cigarettes, to the 'tea cup effect'.

In this book we look back on developments in fluid mechanics in the Netherlands. Burgers' observation, quoted above, raises at least one intriguing question: did Burgers set a trend and can one say that most fluid mechanicists in this country have been 'interpreters' rather than 'discoverers'? It is up to the reader to try to formulate an answer. The topic of this book also leads, naturally, to a related question: has there ever been, or is there today, some aspect of this field of science and engineering which could with some convincing arguments be called 'typically Dutch'? The front cover of this monograph may be seen as a hint toward the author's answer to this question. Although it is an exaggeration, it makes sense to assert that the Dutch were, to a large extent, responsible for the shaping of their country. To achieve their goals, they had to find clever solutions with regard to the inherent threat posed by all the water flowing through and around their country, and they also had to understand its behaviour.

At the Burgers Symposium in June 2018 (the annual event where many of those working in Dutch fluid mechanics come together) the author put up an almost blank poster on which the participants could write down the achievements in Dutch fluid mechanics which they thought have been really unique or important. Among the results were the 'Delta Works' and the making of artificial islands near the coast of Dubai by Dutch contractors, proving that the pride about the Dutch 'fight against the water' is still there. Other achievements written on the poster had to do with "transport phenomena for chemical engineering", "up-scaling", "digitizing PIV", "immersed boundary methods", "convergence CFD non-Newtonian flows", and a few other topics.

↑Flow phenomena for the entertainment of waiting train passengers. During his trips to Delft for consultation in the Library of the TU and the Burgers Archives, the author not only discovered that modern billboards can show moving images but that they also show non-commercial images related to fluid mechanics.

↑Fluid mechanics on Dutch television anno 2018. In March the 'NOS Journaal', the oldest news program of Dutch public broadcasting, reported on simulations which had been carried out by MARIN with a model of the South Korean ferry Sewol. In 2014 this ship had capsized, leading to the dramatic death of more than 300 people. In April 2018 a Dutch daily educational program for children reported on the unique research facilities in a swimming pool in Eindhoven (see chapter 7). During the hot summer of 2018 several broadcasters reported on the 'bubble screen' which Rijkswaterstaat had installed in the Canal between Amsterdam and the river Rhine to prevent salinization (see § 5.3.2). Also in July the Van Heck company, a specialist in pumps since 1964, got international attention by offering its services to save twelve boys from a flooded cave in Thailand. (courtesy of MARIN)

Fluid mechanics in the Netherlands has not really been different from fluid mechanics in other countries. But maybe one rather unique aspect could be mentioned: fluid mechanics has usually been called 'stromingsleer' (flow theory) in this country, so leaving aside the 'mechanics' character of the field. We should add that the term 'stromingsleer' was only used from the 1930s while in Germany 'Strömungslehre' was already fashionable in the 1920s.

Whereas the historiography of (fluid) mechanics has come of age during the last few decades, alas this cannot be said of fluid mechanics in the Netherlands. Fortunately, some of the institutes where flow phenomena have played important roles (like the Hydraulics Laboratory WL and the Aerospace Centre NLR) have initiated the publication of jubilee books in which the history of their activities have been described (more or less extensively). The history of some academic laboratories has also been written down, or the history of some particular fields within fluid mechanics like hydrology and hydraulics. But for most of the topics treated in this volume information could only be found in contemporary sources (of which several are mentioned in the list of references) or in the Burgers Archives in Delft.

The reader may find that a remarkably large part of this book treats researchers and facilities which are related to the University of Delft (THD, later TUD). The author can only admit that since he is a 'product of Delft' and is best known with the Delft archives and libraries it was tempting to put much 'Delft' into this book. But then, there are good reasons to defend this: Delft has been involved in fluid mechanics for much longer than any other university and it has been a source of many and very diverse examples of research, both theoretical and applied.

The author hopes that the content of this book will at least make two things clear to the reader: that many of the current topics in Dutch fluid mechanics have a long history in this country; and that despite the relatively small fluid mechanics community there has always been a large diversity in research and topics. It has not been the intention of the author to give a complete survey of Dutch research in all its aspects; the number of pages didn't allow this anyway. Selection of topics has been based on several criteria. One of them was that there had to be a real Dutch element. Another was: is it something of which Dutch scientists can be proud? Also important was whether enough information and a good image related to the topic could be found.

The author realizes that some readers will be disappointed about the absence of some topics. Some may also be disappointed about the fact that one can find hardly any graphs, tables, formulas, or simulations on these pages. Usually, these kinds of images are only meaningful to a large number of the readers when accompanied by (rather long) explanations. Besides, it would have led to many time-con-

← The author was a PhD student in fluid mechanics in the early 1990s. Here he is discussing some equations related to the so-called vorton method, a rather new 3D vortex method at that time, with which he and others hoped to be able to simulate the vortical coherent structures in turbulent boundary layers. Some 75 years after Burgers started in Delft, fluid mechanics had changed in many aspects, one of them being the use of computers. But in hindsight one can also conclude that the computational hardware and software were still quite primitive as compared to the situation 25 years later (the author was very happy at the time that he could use a '386 PC'). As for the theoretical approach, illustrated by this photo taken in the office of Burgers' 'academic great-grandson' Frans Nieuwstadt, things may not have changed much. This photo is also indirectly pointing to a 'problem' encountered by the author when doing research for this historical account of fluid mechanics: photos showing researchers 'in action' are hard to find.

suming considerations: which graphs, etc., should be used and which not? For the same reason, hardly any references are given to original papers which appeared in scientific or engineering journals.

The reader will also notice that in this book almost all names mentioned are those of professors. You will understand that mentioning all staff members from all groups involved in fluid mechanics would have led to long lists and would have cost a lot of research work. You will also notice that almost all the professors mentioned are not among us anymore. The author has chosen to hold back with the mention of researchers who are still active.

## GUIDELINE FOR THE READER

**The chapters 2–7** in this book can be divided into three segments.

**Chapters 2, 3, and 4** present the history of Dutch fluid mechanics from the 16th century up till the early 1990s.

**Chapter 2** describes some remarkable and sometimes characteristic achievements of the period leading up to 1918. The reader will notice that some kinds of research from the last century had some real pioneers (many) decades before 1918.

**Chapter 3** starts in 1918, which is not only the year in which fluid mechanics got an 'official' status in the academic world (Jan Burgers in Delft) but two other important events for the development of Dutch fluid mechanics took place: the start

of the work which would lead to the Afsluitdijk and several other important hydraulic works related to the Zuiderzee; and the start of the first Dutch aeronautical institute. This chapter ends around 1955, when Burgers emigrated to the US.

**Chapter 4** describes how fluid mechanics developed in the Netherlands after 1955, in the academic world, in some of the institutes, and also in some branches of industry. The 1950s can be seen as the start of a new era. Several important events took place which had a major impact on the development of the field: the start of the Technical University (TU) in Eindhoven; the extension and modernization of the TU in Delft; a strong increase in the work done by Shell on flow phenomena, and the rise of industrial research elsewhere; the rise of numerical fluid mechanics; and the use of computers. This chapter ends with the foundation of the Burgers Centre for Fluid Mechanics in 1992 which meant the 'official' confirmation of the cooperation between the three 'worlds' (the academic, the institutional, and the industrial) and was the starting point for many new interactions.

**Chapters 5 and 6** describe what types of flows have been studied during the last hundred years and how and with what facilities research in fluid mechanics has been done (and how this has changed).

**In chapter 7** 'capita selecta' are presented: some topics which deserve a place in this volume but did not get (enough) attention in the earlier chapters.

© Springer Nature Switzerland AG 2019
F. Alkemade, *A Century of Fluid Mechanics in The Netherlands*, https://doi.org/10.1007/978-3-030-03586-0_2

# 02

## COPING WITH AIR AND WATER IN THE NETHERLANDS BEFORE 1918

The Netherlands: the low countries. Where water easily flows in and is never far away. But also: the (mainly) flat countries. A country with a lot of air above it, with winds blowing and storms passing. It will come as no surprise to anyone that this country has been interested in the behaviour and the flowing of air and water for many ages.

# 2.1 WATER WORKS

## • 2.1.1 IMPOLDERING

It is well known that over the centuries the Dutch have (more or less) constructed their own country by building dikes (or dykes), canals and locks, creating polders, pumping water, etc. One of the first Dutchmen who thought and wrote about the art of constructing dikes and locks was Andries Vierlingh, who died in 1579.

Even in the 19th century one could find several tempestuous lakes ('meren') near Amsterdam such as the Haarlemmermeer, where today one finds the national airport Schiphol. Plans for draining the Haarlemmermeer had already been made in the 18th century but the realization appeared unattainable using windmills alone. The Netherlands was slow in accepting and using steam engines. After the first Newcomen machine had been built and put into operation near Rotterdam in 1776, developments were few, particularly regarding drainage.

In 1847 the Royal Institute of Engineers (KIvI) was founded. Civil servants working for the Ministry (Waterstaat) played a prominent role in this society. The founders were determined to increase the innovative power of Dutch engineering, which had been lacking since the beginning of the 19th century. One of the main points of criticism was the fact that engineers in The Netherlands had failed to integrate the steam engine in their work. Furthermore, the building of dikes, ships, etc., had shown little innovation and a conservative attitude.

From around 1850 more and more people, decision makers and engineers, became convinced of the usefulness of steam pumping stations. The province of Holland, where the number of big steam engines had been low, now started to build the biggest engine in the world, for a draining station called Cruquius. The location was at the border of the Haarlemmermeer, between Amsterdam and Haarlem. From 1850 to 1852 the Cruquius station, together with two somewhat smaller steam-driven pumping stations, managed to drain the Haarlemmermeer completely and gave Holland about 175 square kilometres of new land. Since 1933 the Cruquius is a museum where one can admire the enormous engine with its piston of 3.66 metre diameter and its eight 'arms' which once brought 320,000 litres of water every minute from the lake to the canal around the lake (the Ringvaart). Today the museum shows an instructive and interactive 3D map of The Netherlands where real water is used to point out which parts of the Low Countries would be flooded if we didn't have the dikes, the dunes, and several other constructions which keep us safe.

Models like these for scientific research didn't exist in the Netherlands during the 19th century. From around 1800 the

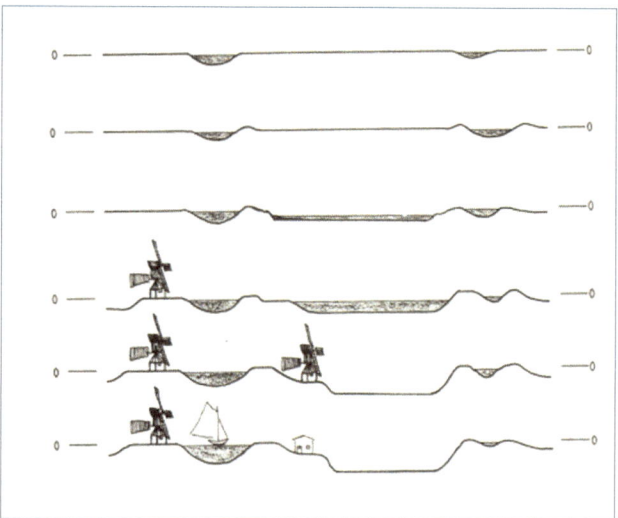

↑The development of the western part of the Netherlands during the last thousand years in a nutshell. (courtesy of and © C.J.M. Tak/www.takarchitecten.nl)

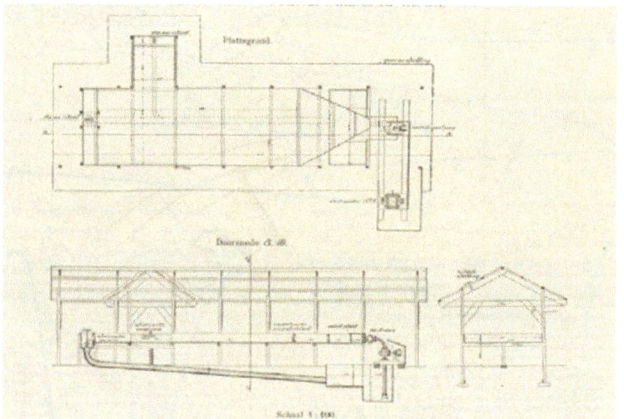

↑Drawings of an 'installation' used to study the mouth of the Nieuwe Waterweg by an engineer of RWS in 1907 and published in the Dutch engineering journal *De Ingenieur.*

French had done some primitive model experiments and by around 1880 these scale models had become more common in France, England, and especially in Germany. However, it was only in the 1890s that Dutch engineers performed the first – primitive - hydraulic model experiments in their own country; usually they had to travel to German laboratories. Among these engineers was the famous Cornelis Lely, a civil engineer who would later rise to become a Government minister.

In the early 1900s a fierce discussion arose about the best way to deepen the Nieuwe Waterweg, the connection between Rotterdam and the North Sea. One critic of the plans

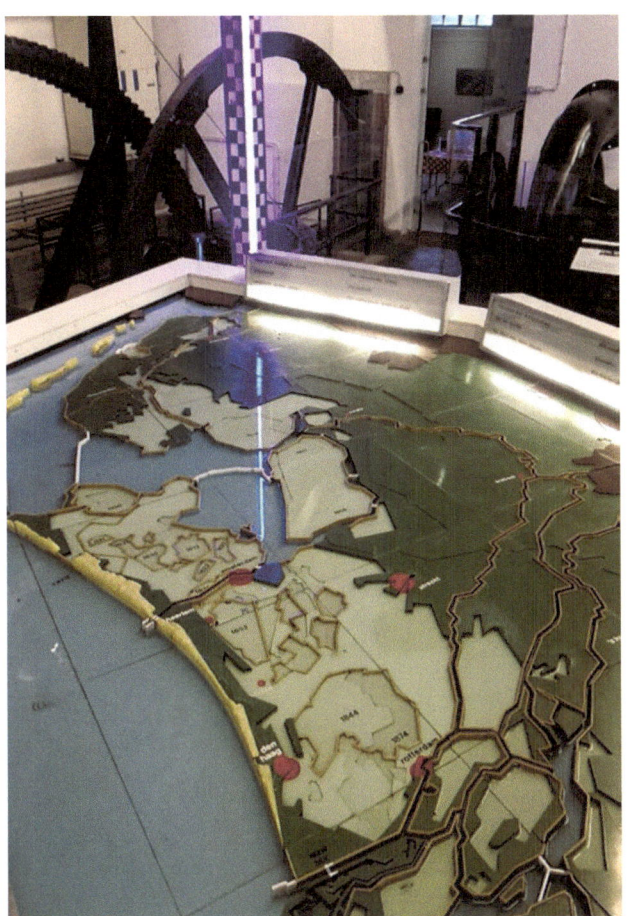

↑Playing with water in the Cruquius museum: what would happen if the Dutch decided to break down their dikes and dunes?

→This statue of the Dutch hydraulic engineer Johannis de Rijke (1842-1913) can be found in Nagoya, Japan. In the period 1872-1903 a small group of Dutch experts in river improvement, harbour design, etc., were asked by the Japanese Government, which had only recently opened up the country, to come and help. The Dutch engineers had already become known for their skills in Asia from their achievements in Indonesia, then a colony of The Netherlands. (photo by Tawashi2006 / en.wikipedia.org/wiki/Johannis_de_Rijke / CC BY-SA 3.0)

↑One of the very first steam engines ('fire machines') in The Netherlands was built in Heemstede (near Haarlem) in 1781. It had to replace the windmill on the Groenendaal estate which had been unable to pump enough irrigation water. (courtesy of Noord-Hollands Archief)

which the Ministry (Rijkswaterstaat-RWS) had proposed, Brandsma, suggested starting a hydraulics laboratory (waterloopkundig laboratorium) at the TH in Delft. In 1907 one of the RWS engineers was ordered to do experiments in a model of the Nieuwe Waterweg that had been erected in the garden of his office. This model was used to gain a better understanding of the flows in the area; unripe cherries from the garden were used to visualize the flows!
But this kind of research was hardly taken seriously in the world of RWS and it lasted only for a year or two. The criticism around RWS increased. To do research in professional model laboratories, the Dutch engineers still had to visit Germany. In the Netherlands the first laboratory would only be put into operation in 1927, in Delft (see chapter 3).

The ultimate water works in The Netherlands were already in the minds of many engineers (and others) long before 1918 but its realisation would occur after 1930: the impoldering of the Zuiderzee, the inland sea that dominated and threatened many a village bordering it. The flooding of parts of Amsterdam in 1916 would be the triggering event.

↓A 'vijzelmolen', a typical Dutch windmill which drove an Archimedes screw (vijzel). This model can be admired inside the windmill called De Adriaan in Haarlem. Around the middle of the 18th century there was a great deal of discussion on the question whether in a drainage mill a water wheel or a screw was more efficient?

↑Windmill 'De Adriaan' in Haarlem was originally built in 1779 to produce cement, paint, and tanbark. It was converted into a tobacco mill some 25 years later. In 1932 it was destroyed by fire and in 2002 it was rebuilt. As is the case with many windmills in Dutch cities, De Adriaan has its sails at a considerable distance from the ground.

## • 2.1.2  MILLING AND PUMPING

Watermills had been in general use from the 12th century. As far as we know, in The Netherlands the first windmills appeared at the beginning of the 13th century. From the beginning of the 15th century these mills were used for transferring water to a higher level. In 1612 fifty windmills had managed to drain an inland sea called the Beemster. The genius behind this huge project was Jan Leeghwater, a famous drainage expert (droogmaker). His plans to drain the Haarlemmermeer would only be realized two centuries later.

## • 2.1.3  DREDGING

Keeping rivers accessible to all kinds of ships has always been important. Especially for The Netherlands: no shipping means no trading. And besides, rivers which are too shallow limit the flow of water from the East to the sea, which can lead to flooding. So, from about 1500 methods were invented and tried to remove mud, sand, gravel, and vegetation from the bottom of rivers and other waters. For the construction of canals in the 19th century, dredging machines also appeared to be indispensable.

Around 1500 dredging was still hard work, being done by hand and shovel (baggerbeugel). Even then the first dredg-

↑Pumping was also necessary in the new installations which arose during the 19th century to take care of water provision for the growing cities. Water towers were part of the water supply chain. This tower is in Dordrecht and was put into operation in 1883 as part of a 'high pressure water pipe' installation along the river Wantij. One of the towers (the original ones have disappeared) contained the chimney of the steam engine that drove the pumps. The river water was pumped to the round tank in the upper part of the tower which could contain 500,000 litres. Since 2007 this is a hotel and restaurant.

↑Dutch pumping inventions: (a) The painter and inventor Jan van der Heyden made important improvements on the fire pump around 1675. This picture shows a somewhat later model which can be admired in the Louwman Museum in The Hague. During the 19th century most of these manually operated machines were replaced by steam-driven versions. (b) The Eckhardt brothers patented this tilted paddle wheel in 1771. It was tested in Amsterdam for the refreshing of the dirty water from the famous city canals. The tests showed that the tilted wheel was not more efficient than the vertical wheel. (courtesy of Nationaal Archief, Den Haag)

ing vessels had appeared. Water ploughs (krabbelaars) were active as early as 1435 in Zeeland province. The harrow or plough loosened the bottom material of a harbour entrance while the tide was ebbing. The ship-like krabbelaar was also moved by the ebb, as it had wings under water which could be spread to catch the currents.

An important new development happened in 1575 when Joost Bilhamer built the first mud mill (moddermolen). Jan van der Heyden (the fire pump man) also invented a dredging mill, characterized by a vertical wheel with compartments. The Holland or Amsterdam moddermolen, which had to keep the river IJ and the harbours of the capital accessible, became well-known.

Thanks to dredging rivers could be made navigable again. Besides the silting of rivers, there was also the fact that the flow of water that entered the country through the upper rivers, was not distributed correctly among the rivers and canals which led to the sea: some got too much water to handle, others too little. During the 18th century it was realized that it would be very helpful if the (average) flow rate in rivers etc. could be measured. It was Christiaan Brunings, originally a German, who in the 1780s invented a flow meter which appeared to give reliable data.

An important new variant on the mud mill was the bucket dredger (emmerbaggermolen), which was used into the 20th century (first with steam engines, later with diesel

↑Models of two types of 17th century 'moddermolens' in Het Scheepvaartmuseum in Amsterdam: the one below uses traction by humans and the other traction by horses. The slurry was drained by means of flat barges or scows.

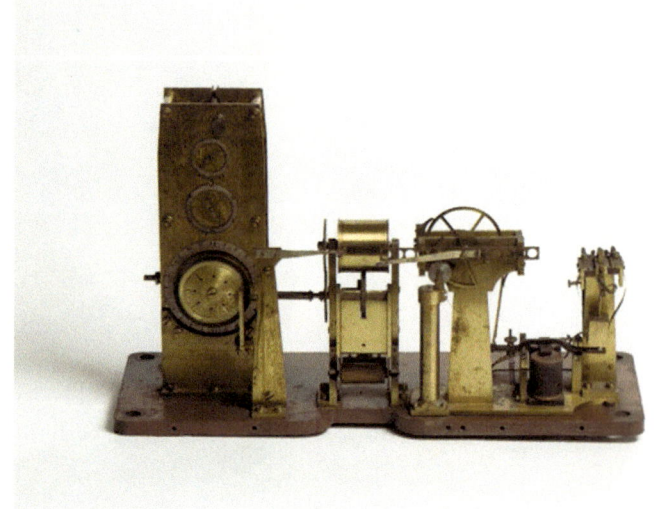

↑An automatic measuring device for wind speeds, called an anemograph. This one was used by the KNMI and built by Olland in Utrecht in 1890. More primitive anemometers for wind had already been built in the 18th century. (courtesy of KNMI / photo by Ed van Rijswijk)

engines). The invention of the piston pump and later the centrifugal pump in the 19th century led to a completely new type of dredging vessel: the trailing suction hopper dredger (sleephopperzuiger), a real autonomous ship which didn't need other boats for discharging. These ships were already used for the construction of the Nieuwe Waterweg, the connection between Rotterdam and the North Sea which was completed in 1872. Soon hereafter Dutch dredgers would be asked to work in many parts of the world.

Fop Smit in Kinderdijk was one of the important builders of dredging vessels (today the company is known as Royal IHC). One of their patents concerned a system of valves with which water could be mixed with sand in a hopper-dredger and then 'ejected' via a pipeline onto the shore where it could be used to make new land.

## C.H.D. BUYS BALLOT (1817-1890)

Christophorus Buys Ballot studied physics and chemistry in Utrecht and defended his PhD thesis in 1844. He started to lecture in mineralogy, geology, theoretical chemistry, mathematics, and astronomy!
Buys Ballot tried to introduce a new theory in physical chemistry but met little enthusiasm in academic circles and he turned to meteorology. At that time this could hardly be called a scientific discipline. In 1848 Buys Ballot started to perform meteorological measurements in a systematic manner. He did this, with a friend, on top of a part of the old stronghold of the city of Utrecht. In 1853 he founded the Sonnenborgh in Utrecht, an observatory for both astronomy and meteorology. Just one year later, he also founded the KNMI at the Sonnenborgh. This place became a place of research and measurement of Europe-wide importance. Buys Ballot was a good organizer and started to use the telegraph to transmit weather forecasts. He was also a promotor of international cooperation and the initiator and first president of the IMO (today called WMO).
The law formulated by him in 1857 and named after him is well known: in the Northern Hemisphere, if a person stands with his back to the wind, the atmospheric pressure is low to the left, high to the right. This is because wind travels counter-clockwise around low-pressure zones in the Northern Hemisphere. The truth is that an American, Ferrel, had actually published this law one year earlier.

←In the Sonnenborgh Museum, Utrecht, in the former building of the KNMI, visitors can learn about (the history of) meteorology. In one of the rooms one finds this bust of Buys Ballot and a weather map of February 3, 1890, the day of his death.

## 2.2 METEOROLOGY AND OCEANOGRAPHY

The draining station Cruquius (section 2.1.1) was named after Nicolaus Cruquius (his Latinised name). Cruquius died a century before the Haarlemmermeer was finally drained but he was one of the first to make plans for this huge project. What Cruquius did achieve is a long series of weather observations and measurements. He was convinced that for proper water management, accurate observations were necessary. Besides temperature, air pressure, and humidity he also recorded the wind speeds, which he derived from the rotation speed of the windmills in his neighbourhood. Thanks to Cruquius, Dutch meteorologists today have a unique, uninterrupted series of data which goes back to 1707.

Cruquius had also pleaded to the Dutch government (so far as this existed in the 18th century) to erect a professional meteorological institute. As with his several propositions with regard to hydraulics, the rulers rejected his idea and it would take till 1854 to realize such an institute, the KNMI. The birth of this Royal Dutch Meteorological Institute (Koninklijk Nederlands Meteorologisch Instituut- KNMI) was largely the work of one man, Buys Ballot.

Buys Ballot thought that meteorology should be more than just publishing weather observations in yearbooks. He was the first to search for coherence between the observations done at different places. Weather charts could then be drawn which led to the possibility of predicting the weather. Making useful weather charts became possible when in 1873 the International Meteorological Organisation was founded.

The KNMI started to publish daily weather maps from 1881. From its foundation it had been housed in the Sonnenborgh observatory (also for astronomy) in the centre of the city of Utrecht. At the end of the 19th century weather measurements became almost impossible in the growing city and the KNMI moved to the nearby village of De Bilt; where it still is today.

During the 19th century, meteorology was to a large extent an empirical science where most of the time was spent doing observations and capturing these in tables and drawings. Only in 1904 the first scientific studies of weather forecasting began to appear (the Norwegian Vilhelm Bjerknes is well-known). It was only after the Great War that weather predictions became more reliable, thanks to the introduction of the 'front theory', i.e., explaining the weather in terms of the interaction of masses of cold and warm air. The role of vortices in the atmosphere became an important object of study.

In 1910 the very first Dutch professor in meteorology was appointed at the University of Utrecht: E. van Everdingen. He had already become director of the KNMI in 1905 and would keep this post up to 1937! During Van Everdingen's professorship the Dutch became especially good in measuring the characteristics of the higher layers of the atmosphere. For this purpose kites and later also airplanes were used.

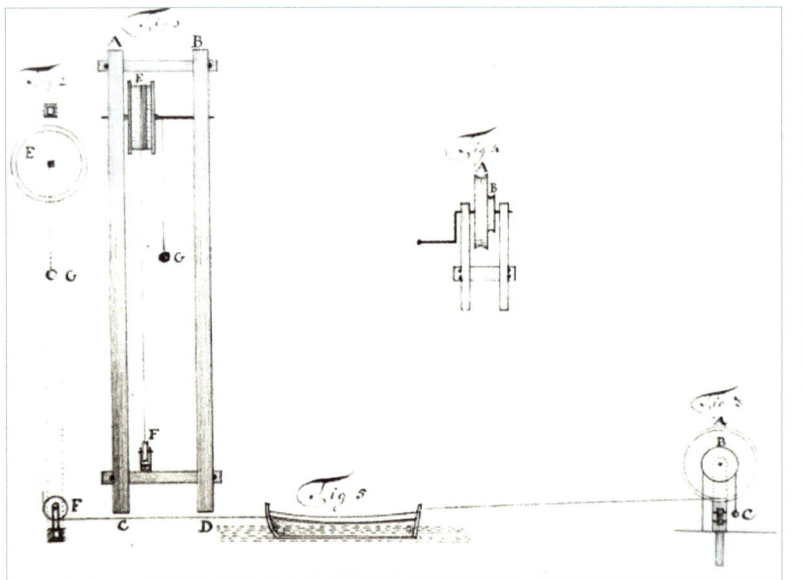

↑Pieter van Zwijndregt never published his towing experiments but he did publish this drawing in 1757. A swinging weight was used to measure the time it took the models to cross a certain distance. (from: De groote Nederlandsche scheepsbouw op een proportionaale regel voor gestelt, 1757)

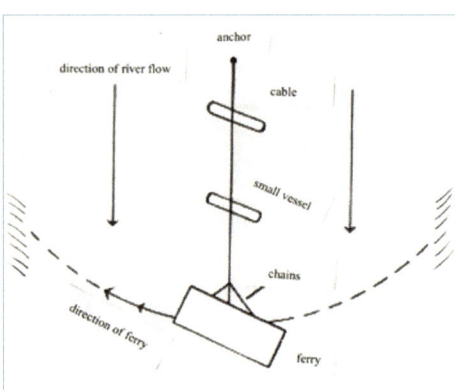

↑The principle behind the gierpont (reaction ferry): the ferry ('pont' in Dutch) is attached with a cable and chains to an anchor. The flow of the river causes a 'lift force' which makes the ferry move along a semi-circle.

# 2.3 ENGINEERING

## • 2.3.1 ENGINEERING ON THE WATER: SHIPS

For many centuries the design of the ships built and used in The Netherlands seem to have been mainly determined by copying from other countries and by optimizing the commercial value of the ships. The 'fluyt' for example, a Dutch design of 1594, got its shape from four requirements: it had to be able to transport a large cargo, it had to be easy to handle and cheap to build and the deck had to be narrow so that less toll had to be paid when passing through De Sont, which linked the Baltic to the North Sea. Speed seems to have been no requirement for a long period and so the flow resistance of the hull got hardly any attention.

An interesting new kind of ship is said to have been invented in The Netherlands around 1600. It was the 'gierpont', a simple ferry boat without sails or engine which is attached to a cable and is moved by the force of the water flowing around it. The gierpont on the river Waal in Nijmegen was in use from 1657 until 1928!

A remarkable Dutch feat of about the same time was the very first navigable submarine which Cornelis Drebbel showed in 1620. It could carry 24 persons, most of them rowers. However, it is still not certain how this vehicle looked and how it was operated. Much better document-ed are the sea-going vessels built in The Netherlands in the 17th and 18th century. But is seems that only one naval constructor took the trouble of doing some research on the performance of different hull lines. In the 1750s Pieter van Zwijndregt towed a series of 1.5 metre-long planks sawed into the shapes of waterlines in a 50 metre long open-water basin. He didn't try to find any formula, he just wanted to know which designs were likely to provide successful ships.

By 1815 the Dutch shipbuilding business had almost complete-ly disappeared. This may explain the fact that the application of iron and steam in Dutch ships lagged behind compared to England and other countries. Some Dutch shipbuilders were good in replicating and optimizing foreign designs. One of them was the already mentioned Fop Smit. In 1852–1853 Smit built the very first completely iron clipper, after an American example. It was very fast for its time, which was partly due to the 'streamlining' of the back of the ship (probably not a result of a scientific approach but of long experience with several other designs).

Smit certainly didn't get help from the Polytechnic School in Delft, since naval engineering was still not one of its depart-ments in the 19th century. Today the students society at Delft is called after William Froude, who did experimental research with model ships in England around 1867, although the results of his work only became available in Delft many decades later. However, there was at least one Dutchman who tried to copy

↑During the 1970s and 1980s most of the Dutch mining heritage disappeared. This picture of two Van Iterson cooling towers was taken just before they were demolished. (courtesy of and © Koninklijke DSM NV)

Froude's work. Bruno Tideman (1834–1883) became Head Engineer Advisor for Shipbuilding (Hoofdingenieur Adviseur voor Scheepsbouw) for the Dutch Navy and started his experiments in 1874 in Amsterdam. He used a scale model of the navy cruiser Atjeh to determine its water resistance and the engine power which would be needed to propel the ship

### • 2.3.2  ENGINEERING UNDER THE GROUND: MINES

Underground coal mining started in Limburg in the 18th century. Two big issues soon became serious obstacles: there was too much water in the mines and too little fresh air. So, water pumps and ventilation systems were necessary. It was only in 1826 that the first steam engine appeared in the mines and it would take another 75 years before the start of serious, large scale mining in the southern part of The Netherlands.
In an article of December 1911 (see also chapter 3) the Delft professor of applied mathematics and mechanics Frederik van Iterson (1877–1957) wrote: "The construction of fans ['ventilatoren' in Dutch] and the development of measuring instruments for air flows, both of high importance for the exploitation of mines, are positively influenced by the progress of aerodynamics. It is sad to see how this branch of science is neglected in the study for mining engineer."
Frederik van Iterson had been appointed professor of Applied Mechanics in Delft in 1910 but left three years later to become

director of the State Mines (Staatsmijnen) in Limburg. His experience with the mechanics of both construction and of air flows surely must have incited him to design a new kind of cooling tower for 'his' mines. The so-called Van Iterson-tower has the typical hyperboloid shape which guarantees both good stability and good draught. The first were put into operation in 1918 at the Emma Mine.

### • 2.3.3  ENGINEERING IN THE AIR: AIRPLANES

During the 19th century, fluid mechanics was primarily water mechanics. Only in meteorology was there some interest in the flow of gases. The first wind tunnel was built around 1870 in the UK but only aircraft pioneers and those concerned with sailing (and ballistics perhaps) were interested in things like lift force. The birth of the airplane prompted an enormous increase in research and activities related to aerodynamics.
In the Netherlands the interest in the new airplanes and in flying started not long after the Wright Brothers and their Flyer became world famous in December 1903. In 1907 the Dutch Society for Aviation (Nederlandsche Vereeniging voor Luchtvaart, NVvL, and from 1912 KNVvL) was founded, mainly due to interest in airships, and in 1909 it started to publish its own magazine. On 27th June 1909 the very first flight demonstration in the Netherlands was given in Brabant by a French/Belgian/Russian aviator, flying in an

↑The first airplane built in The Netherlands: the Van der Burg of 1910–1911. As for cars of that period, streamlining and aerodynamics didn't get much attention, yet... (courtesy of Archief Thijs Postma)

aircraft based on the Wright design.

Up until 1914 the number of Dutch aviation pioneers was small. The most well-known is Anthony Fokker who built his famous Spider (Spin) in 1911 and gave demonstrations with it in Haarlem. However, one year earlier Heinrich van der Burg had built the very first Dutch airplane. The profile of its wings had been 'calculated' by Albert Kapteyn the chairman of the Aviation Section (Vakafdeling voor Aviatiek) of the NVvL. This Section had been founded in 1909 and due to the efforts of Kapteyn an aerodynamics laboratory was erected at the TH in Delft.

In this laboratory it was professor Van Iterson (the inventor of the aforementioned cooling towers) and some students who undertook testing and research in a primitive wind tunnel, made of wood and cardboard. It is not known whether these experiments really helped the very few aircraft

builders in The Netherlands. It was still common practice in the industry, for example at the factory which Fokker had erected in Germany, to base the wing sections on practical experience and on ease of fabrication.

Although Van Iterson left Delft in 1913, the use of the wind tunnel seems to have continued up to the moment that Jan Burgers arrived in Delft as the first professor in aerodynamics. The air battles during the Great War had shown that aircraft would only become more important, not only for the military. Research was soon stimulated by the foundation in 1918 of the Government Service for Aeronautical Studies (Rijks Studiedienst voor de Luchtvaart RSL). The foundation of commercial aviation by KLM in 1919 and the continuation of the Fokker aircraft production in Amsterdam (Fokker had built planes in Germany during the War) gave an enormous boost to aviation.

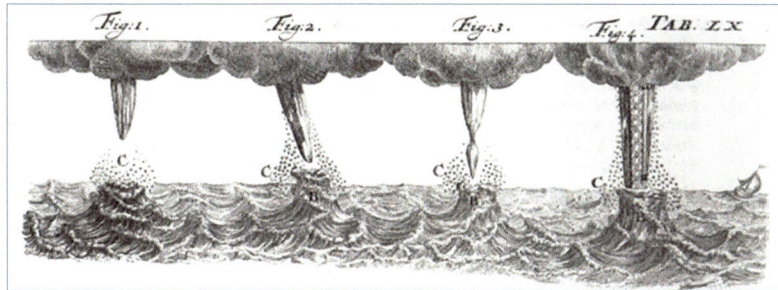

↑As for the physical explanation of phenomena related to fluid mechanics, we find for example an attempt by professor Petrus van Musschenbroek (1692–1761) in Leiden. Not only did he perform meteorological measurements to find out how winds originated, he also tried to understand the behaviour of waterspouts (columnar vortices above a body of water). (from: Introductio ad philosophiam naturalem, vol. II, published by Luchtmans, Leiden in 1762)

↑The very first wind tunnel in The Netherlands, used at the TH Delft by Van Iterson and others around 1913. (courtesy of Stichting Behoud Erfgoed NLR)

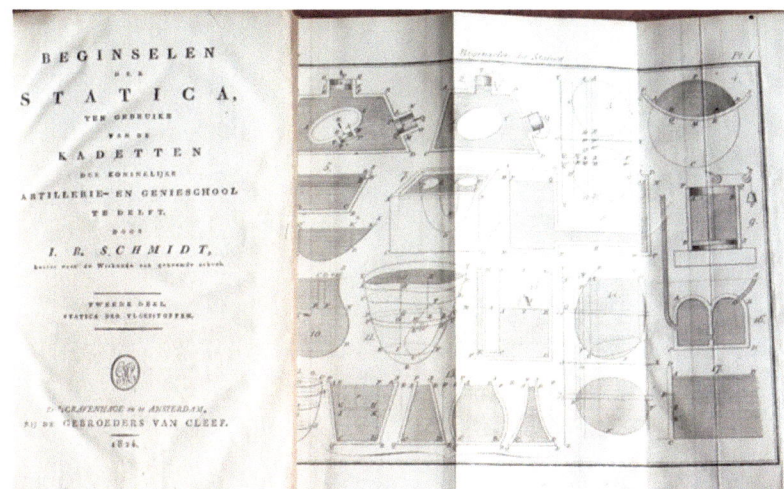

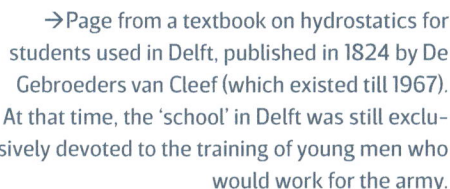

→Page from a textbook on hydrostatics for students used in Delft, published in 1824 by De Gebroeders van Cleef (which existed till 1967). At that time, the 'school' in Delft was still exclusively devoted to the training of young men who would work for the army.

# 2.4 THE SCIENTIFIC APPROACH

The desire to give the flow of fluids a scientific basis dates back a long time. But the road towards a firm theory of fluid mechanics has been tough. In the beginning, still in the 16th century, the lack of mathematical tools was limiting the process. Later on, strange results appeared: the potential flow theory led to the conclusion that bodies moving in fluids could encounter no resistance! For many centuries the distinction between the 'scientific' and the 'engineering' approach to fluid flow was not always clear (and perhaps this distinction was, and is, not very relevant).

One of the very first treatises on hydrostatics published in The Netherlands, and even in the world, was written by Simon Stevin. Stevin showed for example that the pressure of a liquid on a given surface depends on the height of the liquid above it. It was only in 1612 that Galilei wrote a treatise on hydrostatics which really gave this field a thorough foun-dation. In 1628 a student of Galilei, Castelli, published a book which led the foundation for the science of hydraulics and flow measurements.

The most famous and productive Dutch scientist from the 18th century was Christiaan Huygens (1629-1695). Alas, fluid mechanics was never one of his prime topics. But he did discover, from his studies of falling bodies, that the resistance of air varies with the square of the velocity; rather than with the first power as had been currently assumed. He also discovered that the impulse given by a fluid jet on a plate (a situation inspired by water wheels) was proportional to the density of the fluid and to the square of the velocity of the jet. In 1661 Huygens invented a manometer for measuring the pressure in gases.

Five years after Christiaan Huygens' death, Daniel Bernoul-

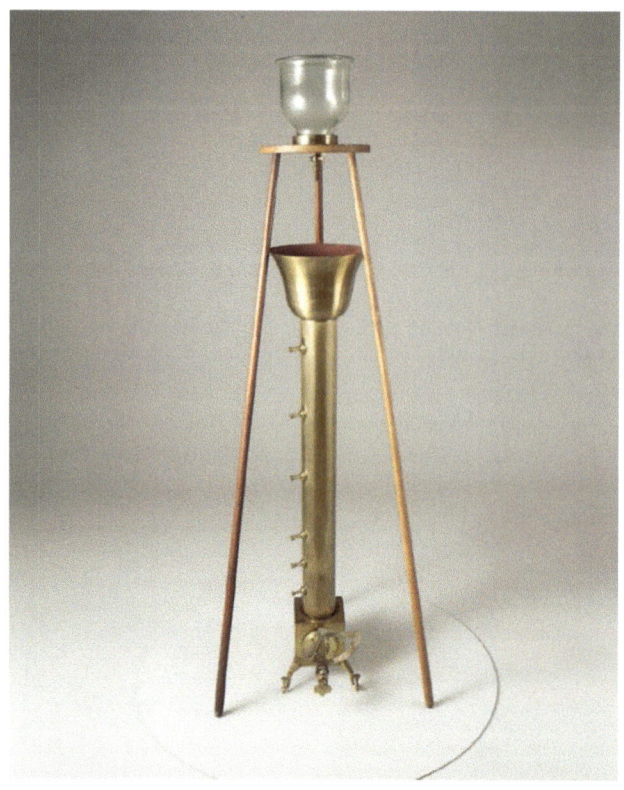

↑At the beginning of the 18th century, the science of hydrostatics had been settled. As for other topics, like sound, magnetism, and static electricity, demonstrations became popular. In Teylers Museum in Haarlem, one of the oldest museums in the world (1784), one can still admire the sometimes ingenious and beautifully made instruments which were used for these demonstrations, and sometimes for further experiments.

(a) This instrument was used to demonstrate the Law of Torricelli: the larger the distance between a hole in the tube and the upper water level, the higher the speed of the water jet. (b) This instrument from 1790 was one of the many to demonstrate the syphon phenomenon: the flow of a liquid from a higher vessel to a lower one where the flow direction is partly upwards but no pump need be used. (c) This instrument of about 1760 was used to demonstrate that water levels in tubes connected to each other are at the same height in all. (d) With this instrument (1790) one could show the influence of air resistance on the rotation of simple propellers. In vacuum they would both rotate at the same speed. (courtesy of Teylers Museum / Trienke van der Spek)

li was born in Groningen, son of the Swiss scientist Johan Bernoulli. He contributed to oceanography and astronomy as well as mathematics but is best known for his work in physics. He established Bernoulli's principle: when in a fluid flow the pressure decreases, the velocity increases. He was also the first to use the idea that a gas consists of tiny particles, thus deriving a number of gas laws (which had already been discovered experimentally). Though born in The Netherlands, most of Bernoulli's work was written in other countries.

Hydrostatics was still the main field in fluid mechanics during the 19th century. At the only engineering school in The Netherlands, in Delft, part of the students trained to become civil engineers. Building dikes and fortifications required knowledge of water pressure. Despite the progress in the mathematical approach of hydrodynamics, for most engineers theory seemed hardly useful: they needed practical knowledge, experience from the past, and simple formulas.
One of the first attempts to explain and mathematize the rise of ground water, an important issue in hydrology, was done by the Dutch homo universalis Pieter Harting (1812–1885) but his explanation was partially wrong and he couldn't formulate the underlying law that the Frenchman Henry Darcy would find some years later, around 1855.

The interest among physicists in fluid mechanics also seems to have been at a low level during the 19th century. Up to the middle of that century the scientific level at Dutch universities was low and innovative research was hardly present. But thanks to a new law on higher education published in 1876 the number of universities grew and their scientific quality increased. The new HBS (Hogere Burgerschool, comparable

to the Grammar school (gymnasium) but without the Greek and Latin and with more facilities for scientific practice) led to an increase in the number of students in the natural sciences. From about 1880 several new physical and chemical laboratories were erected.
The very first thesis in Dutch somehow related to fluid mechanics seems to have been that of Gerrit Jan Michaëlis, titled Over de theorie der beweging van lichamen in vloeistoffen (On the theory of the movement of bodies in liquids). He defended his thesis in Leiden in 1872. A few years later C. Steenhuis got his doctor's degree with his thesis Beschouwingen over de viscositeit van vloeistoffen en gassen (Considerations on the viscosity of liquids and gases; Groningen, 1879). In 1883 the first Dutch thesis dealing with flowing liquids was published: Over de strooming van vloeistoffen door buizen (On the flow of liquids through tubes; M.J.H. Houba, Leiden).
Though the number of Ph.D. students was still small in The Netherlands, two theses related to the same topic in fluid mechanics appeared in the same year, even with the same title (De wervelbeweging). The year was 1888 and both Nicolaas Quint (Amsterdam) and Jan van den Berg (Leiden) had studied the then 'popular' topic of vortical flow. Vorticity had been given a boost by the work by Helmholtz in Germany and the so-called vortex atom theory founded by Lord Kelvin.
In 1888 a certain P. Molenbroek wrote an article titled 'On the theory of liquid jets' (in German) in the Annalen der Physik. Molenbroek published several other highly mathematical papers in the same journal in the years thereafter. His name has been frequently mentioned in later publications by others as one of the inventors of the hodograph transformation, a mathematical 'trick' which allows the transformation of

## S. STEVIN (1548-1620)

Simon Stevin was a very prolific man. He could be named an engineer but also a mathematician or a physicist. He became known for his military methods and inventions, in particular for his plans of defence by a system of sluices controlling the lowlands.
One of his main publications was De Beghinselen des Waterwichts of 1586, which is about hydrostatics. When we read his texts today, it is sometimes hard to follow his way of 'proving' his theorems. And one also gets the feeling that Stevin failed to define the concept of pressure in a way we are used to, as the famous Dutch historian of sciences Dijksterhuis has already noticed in his book on Stevin: "Stevin does not know the idea of a hydrostatic pressure acting at a point of the liquid equally in all directions. The consequence is that Stevin does not succeed in deducing the various subjects treated by him (Archimedes'

principle, hydrostatic paradox, pressure upon an inclined wall) from a single point of view to be brought into relation with static considerations."
Simon Stevin was one of the many who had filed for a patent around 1590 concerning the improvement of mills. He proposed a better working water wheel and an improved method to drive this wheel from the sails. Indeed a few mills were built according to his designs, but the results were quite disappointing. In 1600 he applied for a patent on a dredging boat and a few years later he came up with a huge 'zeilwagen' (land yacht) for Prince Maurits. First a test model was built and then a sailing car for ten passengers. Maurits used the zeilwagen to take guests out for a ride on the beach. It could reach speeds of up to 50 kilometres per hour.

↑ In 2010 a new 'wave gutter' was inaugurated in het public domain of the University of Twente. To celebrate this a team of scientists was asked to make solitary waves in it. Here we see a 'main soliton' together with two 'sub solitons'. Today, solitons are used as e.g. carrier waves in glass fibre optics. (from: Bokhove O, Kalogirou A (2016) Variational Water Wave Modelling: from Continuum to Experiment. In: Bridges T, Groves M, Nicholls D (eds.) Lectures on the Theory of Water Waves. London Mathematical Society Lecture Note Series, 426. Cambridge University Press, Cambridge / courtesy of Onno Bokhove)

nonlinear flow equations into linear equations by introducing new variables. The hodograph method has been thankfully used by scientists working on subsonic and trans-sonic flows. Molenbroek, who remains somewhat of a mystery, was certainly not the first to write about the hodographs but somehow his contribution to its development has been recognized in the field of fluid mechanics.

Between 1888 and 1918, only three Dutch theses related to the theory of fluid mechanics seem to have been published. One would become quite famous: it was on 'long waves', published in 1894 and written by Gustav de Vries. In 1897 the thesis by E.J. Evers appeared: Over de kracht die een vloeistofstroom in sommige gevallen loodrecht op zijne richting op een me-degesleept lichaam uitoefent (On the force exerted in some cases by a flow of liquid perpendicularly to its direction on a body carried along by it). Evers would become a physics and math teacher. In 1901 F.M. Cohen defended his thesis in Amsterdam, entitled De bewegingsvergelijkingen der wrijvende vloeistoffen en de daarbij behoorende calorische vergelijking (The flow equations of liquids with friction and the related caloric equation). This thesis was mainly a (clever) summary of primarily foreign literature. His promotor was Korteweg.

Jan Burgers – who the reader will meet extensively in the next chapters - had also discovered this lack of attention in his country when he started in Delft in 1918. In the Annual Review of Fluid Mechanics of 1975, he remembered:
"Hydrodynamics, although it had a famous history abroad, had not at that time received much attention in The Netherlands. Some outstanding work was done by J.D. Korteweg, professor of mathematics in Amsterdam, after whom the Korteweg-de Vries equation has been named. And there were two fundamental papers by Lorentz. One of the latter treated basic solutions of the Navier-Stokes equations, corresponding to impressed point forces. And the other discussed the theory of turbulent fluid motion in which, among other matters, Lorentz improved Reynolds' estimate for the limit of stability of laminar motion as derived from an energy criterion, by introducing a particular type of elliptic vortex. On the whole, topics governed by nonlinear equations had not yet come into fashion." In 1914 the famous H.A. Lorentz gave a lecture during a meeting of the Koninklijk Instituut van Ingenieurs (Royal Netherlands Society of Engineers, KIVI) on 'hydrodynamical questions' in which he also treated turbulence.
H.A. Lorentz (1853–1928) had been a professor in theoretical

↑In 2011 a remarkable album was discovered in the Netherlands. It appeared to contain dozens of unique very early photographs of people and several Dutch cities. After some research two of them were identified as taken in a backyard of the building of the Royal Academy in Delft, the earliest forerunner of the TU Delft. They are from about 1860 and thus far the only known photos of the Academy. What is even more remarkable: they show experiments related to fluid mechanics. The suggestion has been done that the huge 'barrel' was used to do measurements related to the Law of Torricelli, by the Delft professor Willem Lodewijk Overduyn (1816-1868). (courtesy of coll. Flip Bool - www.found-photography.nl)

physics in Leiden since 1878. One of his first PhD students was Houba, mentioned above. Lorentz's work on the 'point forces', mentioned by Burgers, was published in 1896. In fact, he found a fundamental solution for the flow which today is known as a Stokeslet. Hendrik Lorentz received the Nobel Prize for Physics 1902, together with Pieter Zeeman.

To most scientists active around 1900 physics had more attractive fields to offer than fluid mechanics. Around 1900, atoms and radiation were more interesting for many a young (and older) scientist. Even the rather sensational 'discovery' of the boundary layer by the German scientist Prandtl in 1904 couldn't awaken much interest in fluid mechanics in The Netherlands. Outside Göttingen, Prandtl's non-technical university, the boundary layer theory was largely ignored for almost two decades. "This delayed reception is another manifestation of the gap that separated theory and practice at the beginning of the twentieth century. The same is true for the so-called circulation theory of lift ... introduced ... in the decade before the First World War" (Eckert (2006)).

### D.J. KORTEWEG (1848-1941) AND G. DE VRIES (1866-1934)

Diederik Korteweg started his academic studies at the Polytechnic School, now the Technical University of Delft. Because his disposition for mathematics was stronger than that for technical sciences he switched to the former, but he kept a great interest in the applications of mathematics in physics and mechanics. He wrote a thesis, On the propagation of waves in elastic tubes, under the well-known physicist J.D. van der Waals (Nobel Prize 1910) and defended it in July 1878. The University of Amsterdam had just been granted the right to confer doctorates, and so Korteweg became the first doctor of this university. Three years later Korteweg was appointed at the University of Amsterdam as professor of mathematics, mechanics, and astronomy. In his inaugural address he stressed the importance of mathematical applications in the sciences. Gustav de Vries studied in Amsterdam under Van der Waals, Korteweg, and others. Korteweg also became his thesis advisor. He worked on his thesis while being employed as a teacher at the KMA (Royal Military Academy) in Breda (1892–1893) and at the 'cadettenschool' in Alkmaar (1893–1894). Shortly after his thesis was published in 1894, the main results were made public in the famous Korteweg & De Vries paper titled 'On the change of form of long waves advancing in a rectangular canal and on a new type of long stationary waves'. The equation which has been named after these two men, had actually already been written down by the French scientist Boussinesq some twenty years earlier... Korteweg and De Vries showed that their equation was compatible with the existence of 'cnoidal waves' and 'solitary waves'. Later, it was discovered that other phenomena were also hidden in the KdV equation: long internal waves in a density-stratified ocean, ion acoustic waves in a plasma, and acoustic waves on a crystal lattice. The Institute for Mathematics of the University of Amsterdam is named after Korteweg and De Vries. While Korteweg remained professor in Amsterdam till he retired at the age of 70, De Vries was a high school teacher in Haarlem for the rest of his working life. His grave can be found at a cemetery in Haarlem where Lorentz's grave is also located.

# 03

## THE MAKING OF A 'NEW' BRANCH IN SCIENCE AND ENGINEERING

Though this book takes the year 1918 as 'official' starting point of fluid mechanics in The Netherlands, it would be incorrect to omit the 'pre-work' which has been done in the early 1910s by the already mentioned professor Van Iterson. In December 1911 he

© Springer Nature Switzerland AG 2019
F. Alkemade, *A Century of Fluid Mechanics in The Netherlands*, https://doi.org/10.1007/978-3-030-03586-0_3

wrote an important article in the journal *De Luchtvaart* entitled 'Over de onmisbaarheid van aërodynamische onderzoekingen' ('On the indispensability of aerodynamical investigations').

"Aerodynamics, as a mathematical science, has since long times risen to one of the most beautiful but also one of the most difficult branches of mathematics. ... One can only mourn about the fact that the good Dutch mathematicians leave this field almost completely untrodden. Already the practical aerodynamics has profited largely from the calculations which have been made and we can expect a lot more in the future. But despite all the time and trouble which has been given to mathematical questions, the results which have been obtained from these cannot be compared with the data richness which simple experiments have brought us immediately."

Van Iterson's message was clear: he wanted a laboratory for aerodynamical experiments in Delft, with financial support from the government and other parties and the appointment of a lecturer in this field. He stressed the fact that elsewhere in Europe already several laboratories existed: those of Eiffel in France and of Prandtl in Germany. He gave several arguments for the need of research in aerodynamics: it could make airplanes safer and it could make trains and cars more efficient. He stressed the analogy of results found in aerodynamics and hydrodynamics and the usefulness of aerodynamical experiment for naval engineers (also for the design of torpedoes

# 3.1 JAN BURGERS IN DELFT

At the beginning of the 20th century the Department of Mechanical Engineering (Werktuigbouwkunde) of the TH in Delft was not known for its scientific approach to the topics which belonged to its area of interest: machines, constructions, and production processes. Professors came from industry or had at least strong ties with industrial companies. Differential equations were almost unknown and in most areas of mechanical engineering rules of thumb and experience were still dominant.

Therefore, the appointment of Cor Biezeno, 26 years old, in 1914 as professor of solid mechanics could be called a (small) revolution. The title of Biezeno's inaugural lecture was 'The importance of mathematics as auxiliary science for applied mechanics' and emphasized the need to include applied mathematics and mechanics in the curriculum for engineering education at a sufficiently high academic level. Apparently at that time he had to fight the widespread idea that the engineering profession would need no more than handbook knowledge.

Though by 1914 TH Delft had already reserved a budget for the appointment of a professor, an assistant and a 'servant' in a new division of Mechanical Engineering related to fluid mechanics, it wasn't until May 1917 that a commission was formed to find a second professor in Applied Mechanics. In April 1918 23-year old Jan Burgers, still working on his PhD thesis related to Bohr's atomic theory, was asked to apply for the chair of professor of 'Aerodynamica, Ventilatie en Verwarming' (aerodynamics, ventilation and heating). Thanks to recommendations by his former tutors Hendrik Lorentz and Heike Kamerlingh Onnes (both Nobel Prize winners) and of his thesis supervisor Paul Ehrenfest (successor of Lorentz as physics professor in Leiden) at the end of July Burgers was appointed to the chair of professor of 'Aerodynamica, Hydrodynamica en hare toepassingen' (aerodynamics, hydrodynamics and its applications) at the department of 'Werktuigbouwkunde, Scheepsbouwkunde en Electrotechniek' (Mechanical Engineering, Shipbuilding and Electrical Engineering). From a document from 12th June, 1918 we learn the three persons who had applied for the job had made objections to the original description of the chair: the fields of fluid mechanics and heat transfer were too different to get a manageable professorship! On the 1st of October Burgers started as one of the youngest professors ever in Delft. At that moment he still hadn't obtained his doctor's degree since it was only on 7th November that he defended his thesis in Leiden. On 12th December he gave his opening discourse, titled 'De hydrodynamische druk' (The hydrodynamical pressure) and started to make plans for his own laboratory. At the start of 1919, he only had a room, a chair and a desk.

## LEIDEN

Burgers had studied Physics in Leiden which in the 1910s, as in other periods, didn't include many topics which could be related to fluid mechanics. In 1916 he had been assistant in the already world-famous laboratory of Kamerlingh Onnes where the intriguing properties of helium and other elements near absolute zero were investigated (superconductivity, superfluidity). One of the research themes there was the measurement of 'inwendige wrijving', i.e., viscosity, of fluids like liquid hydrogen. Burgers must have seen these measurements (using a rocking rotating sphere in the liquid) and this may be a reason for the fact that many years later he would become deeply involved in the study of viscosity. At the reading room of the Physics department Burgers had found only a few books on topics related to fluid mechanics. In his unpublished autobiography he would explain: "Ehrenfest had not much feeling for a domain of science which was governed by nonlinear equations, although in 1917 he had directed our attention to a little book by R. Grammel, Die hydrodynamischen Grundlagen des Fluges. [...] This indicated an interesting field for the application of conformal transformation [...] I had sometimes looked at F.W. Lanchester's Aerodynamic theory, but this made the impression of an incomprehensible phantasy."

## INAUGURAL LECTURE

For his opening discourse, he had started to read some of the important literature in fluid mechanics and became acquainted with the recent results by Prandtl, Von Kármán, Eiffel, Kutta, and Joukowsky. He studied Ahlborn's then famous pictures of flow and read extensively in the Jahrbuch der Schiffbautechnischen Gesellschaft. Besides, he discovered the huge amount of investigations which were already taking place on aerofoils and related aeronautical issues, especially those in Germany and Britain.

In his discourse, he mentioned several of the subjects related to his professional area and exposed the difficulty of uniting them in a "righteous" manner. He also realized that, for the moment, he could only base his opinions on the theoretical subjects on which he had read. Only after having acquired long experience, did he think he would be able to assist in industrial problems.

As the main subject of his lecture Burgers had actually chosen the lift on aerofoils which involved the relatively new concept of circulation. The young professor constantly stressed the fact that much remained to be investigated on this issue. He also showed his biological interest by considering the flight of birds.

In 'De hydrodynamische druk' Burgers also carefully exposed his working program which would serve him as a guide during the first decade. After his remark that he would firstly study the theoretical side of his field, he tried to convince his audience (most of them old-fashioned engineers who hardly recognised its value) that theory had to be taken into account in working on problems in his field of research. However, the speaker reassured them, he wouldn't study "the abstract mathematical direction where the problems are only regarded as for the possibility and method of solvability, and where thus the choice of problems is determined by pure mathematical considerations". Instead he would turn to the more physical direction: "study of phenomena, which occur in the flow of fluids and gases, in order to try to solve – by means of the achieved insight – the problems raised by practice". Also, he certainly didn't disregard the meaning of experimental work; he even announced that he had already designed two kinds of experiments.

Burgers realized that the beginning would not be easy, as much had to be started from scratch. In his "retrospective view on hydrodynamics", on the occasion of his departure from Delft in 1955, he remarked: "I was more of a theoretician than an experimenter, and I had a great desire to get to know and to understand many things, and to summarize and categorize these things afterwards; often, the urge for this was bigger than intuition. On the administrative level and as a leader of people, I actually didn't bring along anything; I had to assimilate so many things myself – I was too young and in many senses I have remained like that. I could give people something when they came to me; but I never properly learned how to attract people and to keep them with me". Nevertheless, for the next 37 years he would manage to run a successful laboratory with an enthusiastic staff and to make Delft known as a source of important experimental and theoretical work in fluid mechanics.

## FOUNDING A LABORATORY

Burgers' correspondence with the administrators of the TH on the building of his own laboratory shows that the lack of progress soon started to annoy him. In May 1919 he presented a sketch of the 'provisional' laboratory, being convinced that it should soon be replaced by a real, suitable building somewhere on the outskirts of Delft. He consciously chose for a small laboratory, as he didn't dare to start in a large one. Besides, he wanted to start experimental work as soon as possible. In the letter accompanying his sketches, we read:

"The purpose of this laboratory is the study of all phenomena which take place during the flow of fluids and gases, of the form of the flow, of the pressure on different objects, of friction phenomena, of the transmission of waves, of jets of fluid and gas, etc. In the first place, it will aim at qualitative investigations; furthermore, at the performance of demonstrations for the students in connection with lectures, etc. Naturally, the quantitative investigations should not be neglected; of course, its importance will increase gradually."
As the RSL (see § 3.2) already had some more 'utilitarian' facilities, it was thought that the Delft laboratory should be more concerned with theoretical problems. His early contacts with foreign colleagues helped him in shaping his plans.
The construction of the wooden laboratory was started around May 1920; it was brought into use in the beginning of

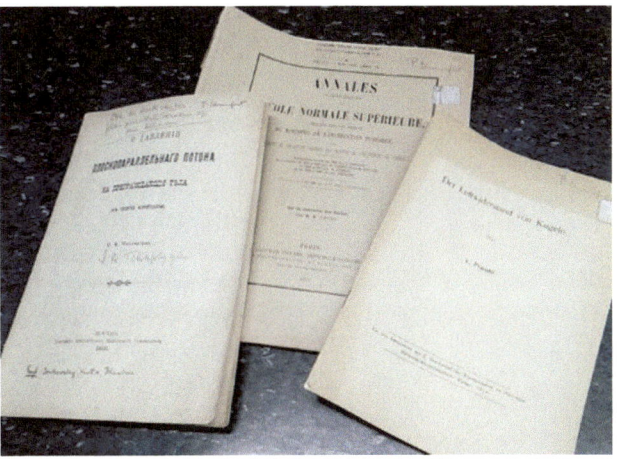

↑During his period as student in Leiden and perhaps also when already in Delft, professor Paul Ehrenfest – the successor of Lorentz in Leiden – regularly sent reprints of articles related to fluid mechanics to his pupil Jan Burgers. Ehrenfest had a broad network, reaching as far as Russia (where he had lived for some years before he came to The Netherlands). It is not well known that Ehrenfest's (hand written) PhD thesis of 1904 was more or less on fluid mechanics. (courtesy of Burgers Archives, TU Delft)

1921. Fifty years later, Burgers still remembered its furnishing:
"One part of the equipment was a small towing tank … of the type used by Ahlborn, whose flow pictures had attracted much attention a few years earlier and had demonstrated the production of vortices of all types of real flow, as opposed to ideal could-be non-viscous flow. Many flow pictures were made and the tank could be used as a welcome demonstration instrument to make my lectures for the students more lively. The other main part of the equipment, built in 1921, was a small wind tunnel of the Eiffel type, with a working section of 4 x 0.8 x 0.8 metres and free return of the air through the room."

By 1929/1930 the equipment was extended with a small wind tunnel of the closed loop type and of dimensions 50 x 50 cm and 2,50 m. The building, however, was still 'provisional'. Though the position of the Laboratory allowed the addition of annexes and although the number of assistants and students would remain rather low till the 1940s, the working conditions for Burgers and his staff steadily deteriorated.

During the first ten years of the Laboratory's existence, the most important member of Burgers' staff would be B.G. van der Hegge Zijnen, who arrived in 1921. He became chief assistant in 1929, although it was only in 1935 that he got a permanent appointment. He would stay until after the Second World War. Bernard van der Hegge Zijnen was a secure and hard worker. It was he who actually put up the whole Laboratory in the 1920s, he who had daily control and he

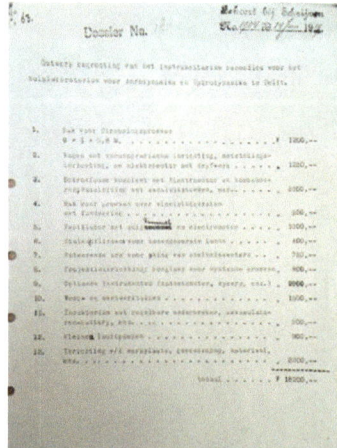

←The very first budget for the Laboratory that Burgers had in mind in 1919. The TH had already found a location near the main building of Mechanical Engineering where it could be erected. In the correspondence of Burgers with the curators of the TH one finds his suggestion to found the laboratory in the still unbuilt polder to the south of Delft where plenty of space would be available to do experiments with gliders. Alas, this polder would only become the location of the TH from the 1950s (and it is still there). Today, the amount of 18,200 guilders would be about €110,000. (courtesy of Burgers Archives, TU Delft)

↑The main experimental room of the Laboratory of Aerodynamics and Hydrodynamics in Delft around 1930. Burgers is on the right, Van der Hegge Zijnen in the middle. Technical assistant Bolsterlee is leaning against the towing tank. The picture shows about a quarter of the room. (courtesy of Burgers Archives, TU Delft)

who was the supervisor of all staff members.
Because of the deafness of Van der Hegge Zijnen and a certain lack of managing skills on the part of Burgers (as he admitted himself in a letter to his assistant), the cooperation between the two men wasn't always smooth...
During the 1930s, the equipment could only slightly be enlarged due to the difficult economic circumstances. Yet, in 1934, a new wind tunnel was built and a device was constructed with which flows could be visualized by means of tobacco smoke. One year later, for the development of the ventilation system of the Maastunnel (see § 3.4) a special shed was built next to the Laboratory for model experiments.
At the end of the 1930s and even during the first years of the War, the situation improved. By 1942, the Laboratory already had four wind tunnels and Burgers still obtained several subsidies from the TH for experimental equipment. However, lack of sufficiently trained people hindered the performing of pure scientific experimental work. War made circumstances even more critical due to measures such as the rationing of electricity. Besides, the building had definitely become too crowded.

In the Burgers Archives in Delft one can find a memorandum listing all the experimental equipment Burgers would like to have in his Laboratory some day:
- a large wind tunnel (diameter 2 meters) with balances;
- for students' practicum: two wind tunnels and a fan;
- a smaller wind tunnel for precise measurements (e.g., in boundary layers);
- an even smaller wind tunnel for flows with very low turbulence;
- a wind tunnel for precise calibrations;
- equipment for the precise calibrations of 'meet-openingen' (openings of measuring devices), both in air and water, also at high speeds;
- equipment for measurements related to viscosity: viscosimeters (also for suspensions) and a centrifuge;
- equipment for studying the behaviour of heavier masses of fluid in less dense masses (e.g., floating particles in air);
- a huge ventilator for one of the wind tunnels for the study of pumps and ventilation;
- a towing tank for both air and water experiments.

After the Second World War general relief was felt and everyone wanted to rebuild the TH as soon as possible. At the Laboratory – from which about 4000 guilders (€23000) worth of machines and books had been stolen during the war according to Burgers' estimate - important changes took place. In 1946 Van der Hegge Zijnen got 'honorary dismissal' after a sick leave due to 'nervosity'. The same year, Leo van de Putte was appointed as lector (the Dutch title for a reader). Van der Hegge Zijnen was followed by a Polish physicist, Lubanski, who unfortunately died only a year

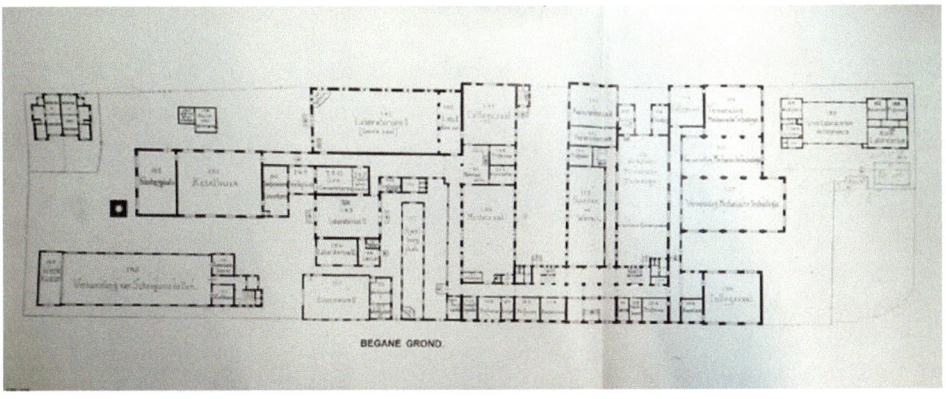

BEGANE GROND.

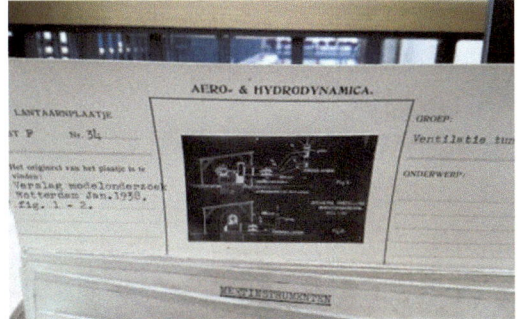

AERO- & HYDRODYNAMICA.

←Burgers' laboratory was built in the early 1920s close to the main building of the Department of Mechanical Engineering and Naval Architecture (today: Maritime Technology). This had the advantage that there was some space around the laboratory where temporary annexes could be erected. This map was used to make a sketch of such an annex where experiments could be done with a fan that was presented by professor Van Iterson to Burgers by the State Mines in 1931. It had been used in Limburg for experiments on ventilation for the mines. (courtesy of Burgers Archives, TU Delft)

←During the 1930s, and maybe even earlier, Burgers made use of a projector ('lantern') and glass slides during his student lectures. Descriptions of these slides are still present in a card slot in the Burgers Archives in Delft. This card shows the slide of the experimental setup at the laboratory with which the ventilation of the traffic tunnel in Rotterdam was investigated. (courtesy of Burgers Archives / TU Delft)

after his appointment. This meant another severe blow for Burgers, who had to manage many of the everyday practical things himself. Besides, the situation at the Laboratory remained primitive (heating, etc.). Fortunately, much of the experimental load was taken up by students and young engineers. By 1948, when Lubanski's successor Betchov arrived from Switzerland, the Laboratory was flourishing again (and became more international).

In March 1949, a second 'hulpgebouw' (auxiliary building) of the Laboratory was officially opened. However, the space problems remained and some students even had to work in the coal-shed! Furthermore, heating was insufficient in wintertime. Officially, the new building was also 'provisional'. Burgers was full of plans but his hope for an extension of the equipment with shock tubes was destroyed when the new Department of Aeronautical Engineering of the TH got priority. However, the frustrating circumstances didn't bother him any longer. He had already made up his mind to leave for the USA. Professor Bert Broer, who had been appointed next to Burgers in 1949, took over the actual management.

### TEACHING

From 1919 Burgers started lecturing. His first lectures were mainly devoted to aerodynamics and aerofoil theory. Fluid mechanics was not taught to students in their first or second year at the THD. To third-year students Burgers gave an elementary introduction to fluid mechanics and to fourth-

year students, who had chosen the theoretical direction, a course on classical fluid mechanics. From about 1935 some of Burgers' lectures were based on his own contribution to the then famous series Aerodynamic theory. (This series was edited by the American William Durand who had invited Burgers since he regarded him as one of the 'big four' in fluid mechanics, together with Theodore von Kármán, Ludwig Prandtl, and G.I. Taylor).

There were also some courses for those interested in aeronautics: 'Motions of air planes', 'Propeller theory', and 'Special topics from the theory of aerofoils'. For fifth-year students, every year some special topic was chosen. These were largely determined by the work, Burgers was doing at the time: 'Wave motion' (around 1928 and 1942); 'Conformal transformations' (around 1929); 'Ventilation' (around 1932); 'Problems from the theory of turbulent flows' (around 1937); 'High speed flows in gases' (around 1946); 'Compressible flows' (around 1949); 'Kinetic gas theory' (around 1949); and 'Gas dynamics' (around 1952).

His lectures for the third- and fourth-year students were never attended by more than about twenty attendants. For general lectures sometimes fifty or more attended. Most students came from his own department of Mechanical Engineering; some of them had chosen an 'aeronautical variant'. Interest from the new Department of Physics, only

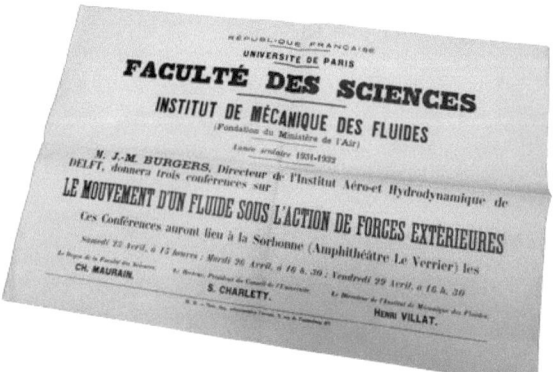

↑Thanks to the 1924 congress in Delft, Burgers and also Van der Hegge Zijnen became known in the whole mechanics community. This led to many contacts, mainly in Europe, and invitations to other meetings. In 1932 Burgers gave lectures at the Sorbonne in Paris. (courtesy of Burgers Archives, TU Delft)

↑Burgers usually felt somewhat ill at ease during official events, like the making of this photo of all participants of a congress on aerodynamics in Rome in 1935. The presence of military people probably only strengthened the unease. Burgers is on the far left. In front of the open door one can see Th. Von Kármán. (courtesy of Burgers Archives, TU Delft)

founded in 1928, remained small before the Second World War.

The number of students who got their doctor's degree under Burgers has been relatively large, about twelve. The first was before the War, as writing a doctoral thesis was rather uncommon in Delft, especially for the 'applied' engineering sciences.

### NATIONAL AND INTERNATIONAL CONTACTS

While the number of professors at the TH with whom Burgers had scientific contact from 1918 remained rather small, he soon corresponded with many outside Delft. His contact with physicists became even more extensive when in 1923 he became chairman of the two-years old Nederlandse Natuurkundige Vereniging (Netherlands' Physical Society, NNV). Burgers was also appointed as member of the Physics Section of the Koninklijke Nederlandse Akademie van Wetenschappen (Royal Netherlands Academy of Arts and Sciences, KNAW). He was a very regular visitor of the monthly meetings in Amsterdam and the contacts he made there meant a lot to him.

Burgers' first scientific journey abroad was to Germany, after Von Kármán had invited him to visit his laboratory in Aachen. He also met several other German physicists and mathematicians, among them Prandtl, whom he met in Göttingen.

In 1922 he was invited, together with Biezeno, to attend a meeting of scientists in fluid mechanics in Innsbruck (Austria). The initiative to this informal meeting came, again, from Von Kármán who had cleverly succeeded in avoiding political conflicts. This meeting not only meant a revival of scientific communication, it was also the first especially devoted to aero- and hydrodynamics. Burgers presented a paper on turbulent resistance, his first attempt in the field of

the statistical theory of turbulence.

After Innsbruck, Von Kármán and others recognized that the entire field of applied mechanics would be served well by regular international conferences. Biezeno and Burgers accepted the challenge to organize the First International Congress of Applied Mechanics in Delft in 1924. However, though Delft was a neutral place, problems arose. Initially Prandtl and some other Germans refused to have any contact with French and Belgian scientists. It was only when the French delegation withdrew (voluntarily), that the conflict faded away. The meeting, for which Biezeno as chairman and Burgers as secretary had to perform an enormous amount of work, was an outstanding success, and it had far-reaching consequences for international contacts and thus for the development of the field. Scientists from 19 countries (among which the USA, USSR, Spain, Turkey, and Egypt) came together to discuss topics like the theory of rupture, motions in rotating fluids, stability of fluid motions, wave motion, and the dynamics of the atmosphere. Of the 214 participants from 21 countries, 105 were Dutch. Next came the Germans with 54 people and the English with only 14. The Delft congress also meant the start of an identity for the field of applied mechanics, which before the Great War had not had any unity at all. This new field in physics had finally found its place at the crossroads.

The 1924 Delft conference was the first in a long list of places around the world where International Congresses for Applied Mechanics have been organized. For many years, Burgers remained a member of the International Congress Committee and was an active participant at most congresses, up until the fourteenth in 1976, again held in Delft. There, he was the only 'survivor' of 1924.

In the immediate post WW II period, Burgers and his colleagues in applied mechanics realized that the organisation of the International Congresses on Applied Mechanics had

## JAN BURGERS OUTSIDE FLUID MECHANICS

It would take many pages to even shortly summarize all the subjects unrelated to fluid mechanics that Burgers did and wrote about and discussed during his life in Delft. At the very least his love of nature should be mentioned: he went on hikes, made drawings of landscapes, collected minerals, and had a sea aquarium in his house. In his philosophical writing one also finds his love for nature when he writes about 'entropy and life functions'. In this alternative life he was very much influenced by the mathematician/philosopher A.N. Whitehead. He also kept a lifelong interest in the philosophical aspects of quantum mechanics; he had seen the birth of this new field from up close in the 1910s.

Burgers always had strong and clear opinions on the position and the correct attitude of scientists. Even in 1918, at the end of his opening discourse in Delft, he told his new colleagues that the common purpose of the staff of the TH had to be the "development of mankind". Directly after the Second World War he became one of the main scientist behind the Studiecentrum voor Maatschappelijke Vraagstukken (Research Center for Social Problems), which – to his disappointment – would soon fade away without having made much change to Dutch society.
Some of Burgers' thoughts on the 'making of science' can be found in a letter of 1953 to a Dutch specialist in the languages and ethnology of the West Indies:
"…for me scientific thinking is not an 'elimination of the I' … On the contrary: all science to me is just a form in which we try to put down suppositions about a coherence of things - reaching farther than poetry, although essentially not absolutely different from it. All science has a meaning for me because of what she tells us, because of the harmony of which she speaks and the joy which she gives us, because of the bond which she can create between people, and of course also because of the power she gives and

which allows one to do more, to have a richer possibility of expressing and to get more freedom in our mind and in all other forms of life."

Special mention needs to be made of a fruitful scientific trip which Burgers made outside the field of fluid mechanics. His work in the Viscosity Committee (see § 5.2) also brought collaboration with his brother Willy. The latter had studied chemistry in Leiden and would become professor of Crystallography at TH Delft in 1940. Together they wrote a chapter in the First Report on Viscosity and Plasticity on the plasticity of crystalline substances. At the end of the 1930s, the Burgers brothers became interested in the theory of dislocations in crystal lattices when Willy had been asked to join discussions on the plastic behaviour of metals. During the war Jan Burgers – and not his brother as many later assumed - discovered the so-called screw-dislocation. This work brought many international contacts and the term 'Burgers vector' got introduced as part of the new type of dislocation.

Burgers visited the USA for the first time in 1930/1931 and was captivated by the people and the landscape. He would only return to the USA in 1949, invited by the Naval Ordnance Laboratory where he admired the new supersonic wind tunnels which had just been built there. About a year later he even stayed at Caltech for six months and the urge to leave The Netherlands must definitely have become very strong by then. But his communist past caused a serious delay in getting permission to emigrate to the USA. In 1955 Jan and Annie Burgers (his second wife) could definitively leave The Netherlands. He became professor at the Institute for Fluid Dynamics and Applied Mathematics at the University of Maryland. He died in Washington, at the age of 86.

to be 'professionalized'. Whereas other branches of science had enjoyed international unions for many years, no such organisation existed as yet for applied mechanics. Partly thanks to the great efforts made by Burgers, in 1946 the International Union of Theoretical and Applied Mechanics (IUTAM) was founded. He also was its first secretary, a position he held until 1952.

Apart from conference visits, Burgers succeeded in making many direct contacts (usually by mail) with scientists all over the world. In the 1920s and 1930s he was visited by the Japanese scientist Tanakadate. In the 1930s, the Belgian scientist M. Biot (later of Harvard) stayed in Delft to work on propellers in pumps. Other important contacts (mainly by letter) included those with Kampé de Fériet (Lille), Körner (Prague), Melvill Jones (Cambridge), Schmidt (Vienna), and Signer (Berne). He visited the Soviet Union several times

in the 1920s and 1930s, giving lectures and admiring the Russian countryside.
As for the German contacts, Prandtl's group would become very important. With Betz, he discussed measuring methods. To Prandtl himself, he sent a lot of reprints and requests for information. In the 1930s Burgers and Prandtl even discussed possibilities of new research in order to avoid duplications in research activities. Their personal relationship did not become very close, although their letters slowly acquired a more personal tone. Of Burgers' British contacts, we have to mention Taylor, who, besides Von Kármán, became one of his closest friends after they had met at the Delft congress. Another long-time British friend was Sydney Goldstein, with whom Burgers would cooperate on a textbook when he had already emigrated to the USA. By then, he also had frequent contact with George Batchelor, founder of the Journal of Fluid Mechanics.

# 3.2 THE RISE OF INSTITUTES RELATED TO FLUID MECHANICS

Some say that the consequences of the First World War for Dutch society (shortages of fuel, construction, and raw materials) caused an awareness among scientists: they could offer the people and industry ideas and advice regarding energy, production, protection against the water, etc.

The fact is that from about 1920 several important scientific institutes were founded (or plans were made for them). All these institutes had ties with Jan Burgers and his Laboratory in some sense. Burgers acted as advisor now and then, got small paid projects for his Laboratory from the institutes and sometimes managed to get cast-off instruments and models from them.

## • 3.2.1 AERONAUTICS: RSL / NLL (NLR)

In December 1918, Burgers was contacted by Emile Wolff (1882-1941), first director of the Rijks Studiedienst voor de Luchtvaart (Government Aeronautical Service, RSL) which had been founded only a few months earlier. Initially, Burgers had rejected membership of the official curatorium (administration) related to the RSL since his pacifistic mind was opposed to the influence of the military at the top of the institute. However, he soon realized that the RSL could help him in getting acquainted with the engineering problems related to aeronautics rather quickly. "A fruitful cooperation developed with the scientific staff of this Institute, which on one hand helped me to see what was done in the world of aeronautics, and on the other hand relieved me of the necessity to move too far into technical matters", he remembered in 1975 in the Annual Review of Fluid Mechanics. The connections with Wolff continued for a long period, during which Burgers acted as an unofficial advisor on aeronautical projects (only in 1937 did he become an official member of the scientific advisory committee). Burgers wrote a paper with the first assistant of the RSL, Pigeaud, and he advised on all kind of problems: wind tunnel construction, corrections for scale effects, and experimental devices and methods. The RSL and the Laboratory borrowed each other's equipment and arranged practical work for students. Besides this Burgers advised the RSL in attracting young engineers who had studied in Delft. In the late 1920s and 1930s, his Russian connections made important Russian literature available to the RSL. In turn, Burgers became acquainted with Prandtl's theory of aspect ratio through the RSL's copies of the Technische Berichte der Flugzeugmeisterei.

Wolff came from the Werkspoor company (see § 3.3.1) and on 1st January 1918 started preparations for the establishment of a laboratory for aeronautics. He became director in April and started with hardly any equipment and only one scientific assistant, electrotechnical engineer Frans Pigeaud. Pigeaud was asked to prepare the construction of the first wind tunnel for the RSL. While the world was still at war, he tried to find any information he could in foreign literature about the tunnels, which were already in use in France and Germany. In February 1918 Pigeaud made a design for the new tunnel (inspired by the tunnel used by Eiffel in Paris) and soon the first orders were placed and the search for an appropriate location was started. This was found at the Marine Etablissement in Amsterdam, a terrain on the city waterfront where the Dutch Navy had been since about 1800. The aerodynamical balance and some other instruments for the wind tunnel came from the pre-war Delft wind tunnel (see § 2.3.3). On 5th April, 1919 the RSL was officially opened.

The range of activities of the RSL soon grew. There was not only aerodynamical research, but material tests, engine tests, and flight tests were also performed. A close relationship with the Fokker airplane factory, the Army, and some ministries started. In 1920 the RSL came under the supervision of the Ministry of Public Works (Waterstaat). There the civil servants and the minister discovered that doing aeronautical research was a costly business and in 1926 it was decided to abolish the RSL. A Special State Committee advised keeping the RSL alive but to move part of the research departments to the TH in Delft. However, Delft refused to accept the offer and the RSL remained in Amsterdam. From 1929 the situation became (more or less) stable and the number of staff members increased. One of the Wolff's first collaborators was Henk van der Maas, who in 1940 became the very first professor in aeronautical engineering in The Netherlands (see § 4.1.1).

The first RSL wind tunnel was used for twenty years but in the 1930s it became clear that as airplanes became faster, the speeds in wind tunnels (and the Reynolds numbers) should increase too. In 1940 two new low-speed wind tunnels became operational in a new building on the outskirts of Amsterdam. In 1937 the name of the institute had been changed to Nationaal Luchtvaart Laboratorium (National Aeronautics Laboratory, NLL) and it had become an independent non-profit foundation. During the war the NLL continued its activities and was not directly involved in research related to the war effort. Some of the aerodynamical research activities were related to the calculation of lift distribution of wing planforms, flutter and vibration of airplanes, and the dynamics of all-wing airplanes. Besides fluid mechanics also the mechanics of materials and airplane behaviour were topics of research at RSL/NLL from the beginning.

←For several years volunteers set up and managed a museum about the RSL/NLL/NLR in the NLR building in Amsterdam. One of the objects visitors could admire there was this model of the screw for the two wind tunnels which were put into operation in 1940. It was designed by Burgers and built by the Jaffa company in Utrecht, whose director, D. Dresden, was also professor at the TH Delft.

←The NLL around 1950. On the left one sees the high building containing the power station. To its right the impressive high-speed wind tunnel which was officially opened in 1960. (courtesy of Stichting Behoud Erfgoed NLR)

Wolff, who was Jewish, had already resigned in 1940 after becoming seriously ill and died in 1941, just before the massive persecution of the Jewish population started.

After the war the NLL, encouraged by the Government's intention to promote the design and development of aircraft in The Netherlands, prepared new plans, including the construction of a transonic and a supersonic wind tunnel. However, around 1950 the Dutch Government seemed to withdraw its (financial) support of all NLL plans and the future looked dark. The report of a 'heavy' committee, in which Van der Maas took part, saved the future of the NLL and led to an amazing combination of three jobs for Van der Maas: head of the NLL, head of the Department of Aeronautical Engineering of the TH Delft, and head of the Netherlands Institute for Aircraft Development (NIV)! In 1955, more than ten years after the war, the construction of the High-Speed Wind Tunnel of the NLL could finally be started (see also § 6.2).

## • 3.2.2 HYDRAULICS: WL (DELFT HYDRAULICS, DELTARES)

In the 1910s the Dutch government had decided to dam the Zuiderzee, the inland sea in the heart of the country, by building a 'dike' (the Afsluitdijk or Enclosure Dam). A committee was appointed in 1918 for this purpose and its achievements can be regarded as the glorious swan song of its chairman, the old and widely-respected H.A. Lorentz. The State Committee for the Zuiderzeewerken (SZ) had as its primary task to investigate the consequences of the Afsluitdijk. Its final report on this matter appeared in 1926, but it remained thereafter occupied with several related questions. Many a Dutch scientist and engineer became involved in the challenging projects of the Committee (see § 3.4.1 for more details).

↑Research in the cellar. It is not known on which project this staff member is working around 1930 but one of the first research jobs for the WL was related to the Wilhelminahaven in Vlaardingen, near Rotterdam. With this model, in the cellar of the building of the Department of Weg- en Waterbouwkunde, one tried to get insight into accretion in this harbour. (courtesy of TU Delft / photo by Fotografische Dienst via Beeldbank / CC BY)

↑The model of the northern part of the delta area of the WL is seen here during its construction, around 1948, in the open air next to the WL building. Local residents had a nice view of the remarkable contraption. (courtesy of Deltares)

An important role in the project was played by the Water-bouwkundig Laboratory (WL), a national research institute on hydraulics and civil engineering (later its name would become Waterloopkundig Laboratory; 'bouw' in Dutch means building, 'loop' means course).

The story of the birth of the WL is rather curious. In 1919 the Dutch government had already decided to erect a laboratory for hydraulic model experiments. However, the engineers at Rijkswaterstaat were sceptical of the necessity of such a facility, and it was not until 1926, after many discussions, that a provisional laboratory was opened in one of the cellars of the Department of Weg- en Waterbouwkunde (now: Civil Engineering) of the TH Delft. The new facility was given a year to prove its usefulness and viability, and it did so without any problems. In 1927 the WL was a fact and as its first director Thijsse was appointed. Soon, the engineers at RWS had to admit that the WL was indeed an important step forward, e.g., in the design of new locks for rivers. In 1933 the WL got its definitive recognition, and finances, and was turned into a foundation. Most members of the board were appointed by two ministries. The WL would have its premises in the TH Delft, but it would not be part of it. In this way, such was the idea, it would remain an independent scientific research facility.

In some respects the WL was a unique facility in the world. Thijsse had started research on wind waves and he realized that he needed 'real' waves for this: not waves generated by a moving object in the water but waves generated by artificial wind. In 1935 the world's first 'windgoot' (wind gutter) was put into operation. (Though some sources indicate that a combination of a wind tunnel and a 'water gutter' had already been in use at the Laboratory for Experimental Geology at the University of Leiden some years before.)

RWS had been a rather conservative institution for many decades, but during the 1920s things changed. In 1930 the Studiedienst van de Zeearmen, Benedenrivieren en Kusten (SZBK = Study Department of Sea Arms, Lower Rivers and Coasts) was set up by RWS, another sign that the scientific approach to hydraulic engineering was gaining acceptance. The SZBK had to study the important delta area of the southern part of the Dutch coastline, an area for which knowledge was still largely lacking. It studied the flows there, the transport of sediment and sand due to tidal movements, and the mixing of salt and fresh water. To this end new measuring instruments, e.g., for investigating flows under water, were acquired and sometimes adapted. It soon became clear that a mathematical approach to the delta area would be even more difficult than that for the Zuiderzee project. A model of the area was very much needed, but the WL lacked space for this new project. After the war part of the space problems could be solved by using

## J.TH. THIJSSE (1893-1984)

Jo Thijsse had a famous father: the biologist and conservationist Jac. P. Thijsse, the man who would teach the Dutch to love nature in their country. His son also loved nature but chose to become an engineer. He studied Civil Engineering in Delft and had just finished his studies when, in 1918, the Lorentz Committee started its activities; and it was looking for a young engineer interested in hydraulics. So it happened that Thijsse became involved in the biggest ever project in The Netherlands: the transformation of the Zuiderzee into the IJsselmeer by the building of the Afsluitdijk. Thijsse was made responsible for the measurements which had to be done in the North Sea. He also got involved in helping Lorentz with the formulation of calculation models and the execution of the tedious calculations themselves.

In 1920 Thijsse also became employed by the Dienst der Zuiderzeewerken. During the 1920s he was a 'liaison officer' between the Dutch and the model basins in Karlsruhe, Germany. The hydraulics laboratory in Karlsruhe had been founded by Theodor Rehbock, who had been born in Amsterdam but trained in Germany.

This was how Thijsse became interested in the usefulness and necessity of model research in basins and he became the expert in The Netherlands. Therefore, it came as no surprise that he became the first director of the WL in 1927. Thijsse was also the initiator of the 'dependance' (annexe) of the WL in De Voorst, which was put into operation in 1951.

Thijsse started to give lectures in 'theoretical and practical hydraulics' in Delft in 1936 but it was only in 1946 that he became an ordinary professor. At the same time, he remained director of the WL and advisor for the Dienst der Zuiderzeewerken. In the 1930s Thijsse tried to convince the faculty of Civil Engineering in Delft to insert a practicum (internship) for students in the WL but this appeared to be not so easy. Only in 1939 did the first students carry out measurements and calculations in the WL. Very few

students could do their graduation studies with Thijsse, since he regarded hydraulics not as an independent field of research but as one helpful for other fields of engineering. One can conclude that Thijsse never tried to introduce the scientific approach in the faculty of Civil Engineering. Only a few have become a doctor under his guidance. Thijsse was also involved in the foundation of the International Course in Hydraulic Engineering, today called the IHE Institute for Water Education. Here students from (mostly) developing countries can become acquainted with the Dutch experiences in hydraulics. For many years Thijsse was secretary of the International Association for Hydraulic Research (IAHR; which today stands for International Association for Hydro-Environment Engineering and Research).

One of his mottos was: It is the art of the engineer to draw sufficient conclusions from insufficient data.

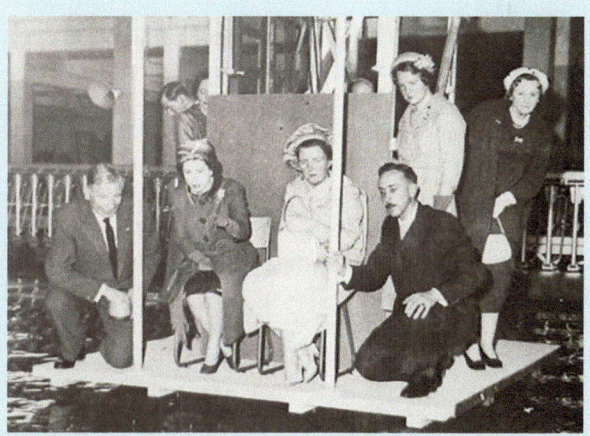

↑Thijsse was regularly a host to important guests from around the world. Queen Juliana loved to visit the WL with her guests. She is seen here with her colleague from Great Britain during a visit in 1958. Jo Thijsse is explaining on the left. (photo from: Dirkzwager (1977))

the terrain next to the WL building. There from about 1948 a model of the northern part of the delta area was built, from cement mortar and plywood moulds.

The WL under Thijsse initiated research in several fields: tidal flows, storm surges, high water waves on rivers, groundwater flows, translation waves in canals, transport of sediments, etc. Most research was done at the request of clients (which increasingly came from abroad). Around 1952 Thijsse noticed that there was hardly time left for new, innovative, demand-independent research. His plans for a new policy at the WL would soon be 'disturbed' by a long range of studies which were ordered by the Deltacommissie (Delta Commission). This commission had been installed

only 17 days after the nightmare which some had already predicted: the North Sea Flood of January–February, 1953. The extension of the harbour at Rotterdam (e.g., Europoort area) would also lead to an increase in the activities and the number of personnel at the WL in the 1950s and 1960s. At the same time the more 'scientific' and theoretical approach of hydraulic phenomena within the WL took shape. Mathematicians were hired, PhD theses were published, and the computer made its entrance.

A remarkable extension of the WL took place from 1952 in the Noordoostpolder, a new polder which had been part of the Zuiderzee/IJsselmeer before 1942. There, in the wide-open countryside, the WL location 'De Voorst' was literally

↑The construction of the NSP caused a remarkable sight for passers-by. Here we see the deep-water tank which originally had a length of 160 meters, but which was then extended to 252 meters in 1951. For many years the NSP, or NSMB, and now MARIN, has been associated with the city of Wageningen (well known for its Agricultural University but not for shipbuilding). Today MARIN also has an office in nearby Ede and in Houston, USA. (courtesy of MARIN)

dug out. In total 35 scale models of sea arms and harbours were created and used to observe the behaviour of the water when dams and other constructions were inserted (see also § 6.2).

### • 3.2.3  SHIP DESIGN: NSP / NSMB (MARIN)

Tideman (see § 2.3.1) had used a model basin around 1880 mainly for the investigation of the resistance of ships but for many years after him nothing happened in this field. It was only used to calibrate flow meters from time to time. In the 1910s the curators of the TH in Delft considered the building of a ship model basin but concluded that it would be too expensive if it would only be used for educational purposes.

Only in the 1920s the so-called towing tank ('sleeptank' in Dutch) became used for investigations concerning hydro-dynamic loads and ship hull excitations. When in 1922, after the great economic depression, the ship building industry recovered, the first Dutch towing tank was built. This project was realised thanks to contributions from several parties. The Stoomvaart Maatschappij Nederland, the Royal Rotterdam Lloyd, the Royal Packet Boat Service and the Nederlands-Indische Tankstoomboot-Maatschappij (which exploited steamships in the Dutch East Indies) were prepared to pay half the foundation costs. The remaining 350,000 Dutch guilders were paid by the Dutch govern-ment. This was the start of the Nederlandsch Scheepsbou-wkundig Proefstation (The Netherlands Shipbuilding Test Station, NSP), the basis of today's MARIN. In international contacts and publications in English the institute was called the Netherlands Ship Model Basin (NSMB).

Although an article on the construction of basins for model experiments had already appeared in De Ingenieur in 1918, plans to found a serious research institute for ship research

with basins were only made at the end of the 1920s. In 1930 Burgers was appointed as member of the Commissie van Bijstand (Commission of Assistance) of the NSP Foundation. This foundation was set up in 1929 by the four shipping companies mentioned above and the Dutch State. The construction of the building and the facilities of the NSP in Wageningen were completed in 1932, under the man-agement of professor Lou Troost (1895-1966) who was the director of the NSP for twenty years, before he emigrated in 1952 to the USA, to become dean at MIT.

In April 1932 the test station was put into operation. After the 160m long Towing Tank (which later became known as the Deepwater Tank) was filled with water, the first towing tests were carried out. On 9th May 1932 the NSP was offi-cially opened, including a facility where ship models could be made. At the time there were no permanent employees. In that first year, 1932, the number of orders booked was far higher than expected and the exploitation costs of about 30,000 guilders (€275,000) were to a large extent recuper-ated. This favourable result was continued in the following years and foreign orders also started to come in. The star of the NSP rose quickly and in 1933 it took the initiative to organize the first International Conference of Tank Superin-tendents in The Hague. Burgers was one of the participants.

By 1938 the number of employees had grown to 36, which required an extension of the number of offices. The demand for model tests grew steadily, not only on the behaviour of ships (rolling, pitching, etc.) and water resistance but also on propeller cavitation. In 1941 a large cavitation tunnel was constructed, at that time the largest in the world.

The laboratory suffered severely from the war. Research activities were at a low level but this gave the staff the opportunity of writing a comprehensive text-book which

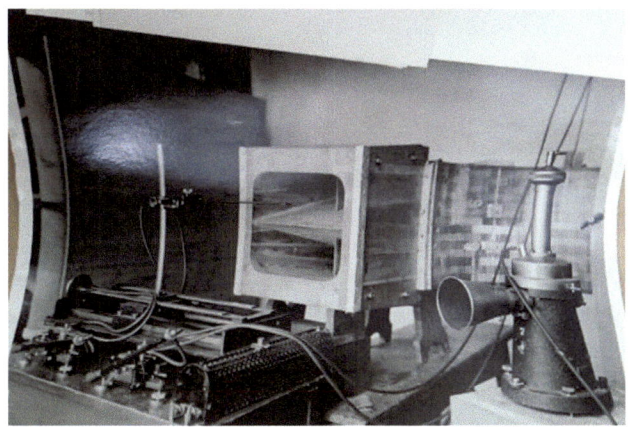 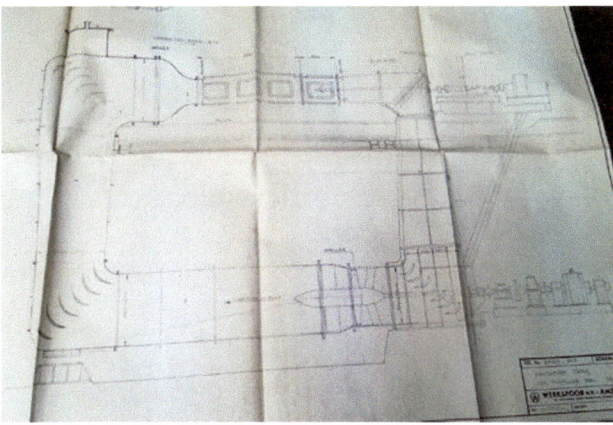

↑For the design of the cavitation tunnel of the NSP, which was to be built by the Werkspoor company, Burgers' advice was asked for in the period 1939–1943. In his laboratory a model of parts of the tunnel were tested (photo). (courtesy of Burgers Archives, TU Delft)

was translated in 1948 into English and became well-known: Resistance, propulsion and steering of ships – a manual for designing hull forms, propellers and rudders. After a forced evacuation in October 1944, looting by the Germans meant that most of the important machines were taken to Germany. Some of these were traced and recovered after the Liberation, but the chaos was so complete that operations could only be resumed in March 1946, and then only partially.

In 1952 the NSP was completely recovered. In a booklet on its activities, the 'superintendent' Wim van Lammeren (1908–1992) wrote: "The object of the experimental station is to promote efficient ship designing by means of scientific research in the field of hydrodynamic problems connected with it. This end is pursued chiefly by carrying out and analysing experiments on a reduced scale model with a view to ascertaining the most economical shape of ships' hulls and their propellers." That same year the Deepwater Tank was extended to a length of 252m and the design department was increased considerably. In the following years the number of basins was further increased (see also § 4.2 and § 6.2).

# 3.3 FLUID MECHANICS IN INDUSTRY

During the early years of the twentieth century, a number of Dutch companies started to grow at an impressive rate, some of them to become real multinationals. Some became involved in various fields of engineering and became almost household names. Others started to develop their own laboratories for research and development.
During his professorship Jan Burgers was regularly consulted by various kinds of industrial companies, big and small, even during the war period 1940–1945. The work that he did for

the engineering company Werkspoor especially established his reputation as a reliable and thorough partner in solving engineering problems related to fluid mechanics. His policy regarding commercial consultancy work is not known, but it is evident that he didn't reject it. On the other hand, he never became a commissioner for any company, as some of his colleague professors in Mechanical Engineering did. It could be that Burgers saw his support in solving industrial problems as a means to help society in general. And when one reads the constant complaints about lack of money in the annual reports of the Laboratory, especially during the 1930s, one can even better understand his willingness to accept paid jobs from industry. Maybe he also felt flattered. As the Machinefabriek J.J. Kremer in Groningen wrote to him in August 1945: "The experiments which we have done so far have cost us so much money already that we have decided to consult the highest authority in this field."

It occurred quite often that companies, usually small ones, came up with questions that were poorly defined and did not attest to a great deal of insight into the flow theory. To Burgers was then the task of pointing this out; but he almost always tried to help them.
This he did with a request from the company Nepaton N.V. in The Hague, in the years 1942 and 1943. The peculiarity in this case was that it concerned a clear war application: the air resistance of projectiles at high speeds. Burgers apparently did not have any moral objections to cooperating. He may have thought that he should be able to participate in all possible ways in the fight against the Occupier. Nepaton wanted to release molecular kinetic theory on the case, but Burgers soon saw that this approach was probably too fundamental. Moreover, the memorandum that he received from the firm was vague and (more or less) a summary of some literature written by someone who had little knowledge; in fact it was unusable. Burgers therefore replied in

47

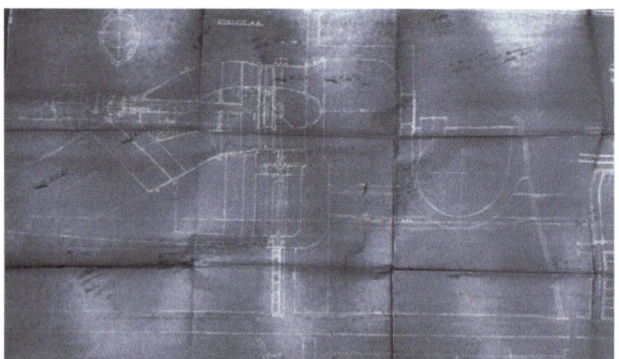

↑Another document from the Burgers Archives in Delft. To discuss a problem related to the pumping station near the Linge, Burgers studied the blueprint of the installation drawn by the people at Werkspoor. His handwriting is on the drawing. (courtesy of Burgers Archives / TU Delft)

↑In 1952 Werkspoor issued a booklet on the occasion of its 125 years of existence. It was a showcase of all their products of past and present. These are the fans for the ventilation system of the mines in Heerlen, Limburg in 1933. (TU Delft Library collection)

a letter: "there cannot be any remuneration for me under these circumstances, although I would like to be helpful in literature studies, as I have said." To which he added: "May I end by asking attention for one particular aspect: if others were to ask my judgement on the thoughts you have developed, I cannot give any other judgement at this time than those formulated in the preceding pages [of the letter]. Even if, as one can understand, I would express myself very succinctly, it would be possible for others to get a certain opinion from my words. I would therefore want to ask you not to mention my name in any discussions with third parties."

### • 3.3.1 WERKSPOOR

The precursor of the company that would become known as Werkspoor had already grown into an important player in the world of machine construction during the 19th century. Around 1900 it built trains and ships and it soon went on to become known for its diesel engines and bridges. Werkspoor also became involved in the construction of pumping machinery, and as we have seen in § 3.2.1 they built the very first wind tunnel for the RSL.

Burgers' first contact with Werkspoor took place in the early 1920s, in the context of the Zuiderzeewerken (see § 3.4). Part of this project was the draining of the inland lake called the Wieringermeer, for which two pumping stations ('gemaal' in Dutch) were built in the 1920s. The station at Medemblik had three identical centrifugal pumps with vertical axes.

In order to find the most favourable shapes for the pump channel and impeller, extensive model experiments were performed at the Waterbouwkundig Laboratorium in Delft. But the scientists working on the Zuiderzee also decided to take the theoretical road. Around 1923 they asked Burgers to set up a theoretical approach of the flows around the blades of the impellers. In his reaction Burgers proposed to 'translate' the job into: "the formulation of a calculation method which makes it possible to investigate some particularities of the flow in certain impellers along a theoretical road". One doubts whether Burgers realised at that time how tough this job would be. Five years later Burgers and Van der Hegge Zijnen were able to publish their report for Werkspoor, after years of tedious calculations (see also § 6.3). In later years Burgers would again be asked to design the ducts for pumping stations.

During the Second World War Burgers was again involved in a problem related to a pumping station, this time near the river Linge. Here he had to advise Werkspoor on the channel through which the water had to flow towards the screw pump in the pumping station. To solve the problem a wooden model of the channel was built in Burgers' laboratory.

In the pre-war period, the coal mines in the south of The Netherlands were managed by a company called Staatsmijnen (State Mines; the name lives on in the chemical company DSM = Dutch State Mines). Contact between Burgers and Staatsmijnen was arranged by Van Iterson, the pioneer in aerodynamical research mentioned in chapter 2, who was one of its directors, and by Th. W. Theunissen, who was then head engineer of

←↑The most important competitor in the field of pumping machinery of Werkspoor was Stork. Stork built the equipment for the Kolff pumping station near Gorinchem which was opened in November 1945 and was at that time the largest gemaal in The Netherlands. It still contains the three original Stork pumps today (diameter of the impeller almost four meters); the diesel engines have been replaced. A remarkable fact but probably easy to explain: Burgers was never asked to do research by Stork, the other main builder of pumps in The Netherlands. This despite the fact that his first PhD student, Borren, was director of Stork for many years. In 1954 Stork would merge with Werkspoor.

Werkspoor. Both men asked Burgers in the late 1920s to solve problems related to the ventilation of some of the mines. In the early 1930s, a model of the ventilation system of one of the mines was built at the Delft Laboratory which clearly showed the causes of the irregularities which had been encountered. Taking these results into account, Theunissen soon found a simple remedy for the problem. This must have strengthened Burgers' esteem for the practical aspects of fluid mechanics. During the Second World War, Burgers worked on problems related to the cooling towers of the mines and had discussions about the problem of the coal dust which was transported by the ventilators of the mines. After the War, in 1947, Burgers was even asked to become official advisor to Staatsmijnen on aerodynamical and hydrodynamical issues. However, he rejected the offer since he was occupied with too many other activities and for fear of the bureaucratic involvements.

In the 1930s Burgers also had contact with Werkspoor in the construction of the very first traffic tunnel in The Netherlands. In 1933 the city of Rotterdam had decided to build a tunnel under the river Maas, the Maastunnel. Burgers was asked by Theunissen to design the ventilation system and to test a model in his Laboratory. Work on the model started in the Laboratory in 1934 and lasted for nearly five years (see § 3.4).

## • 3.3.2 SHELL

In 1890 the Koninklijke Nederlandsche Maatschappij tot Exploitatie van Petroleumbronnen in Nederlandsch-Indië was founded to explore oil fields in Indonesia, then a Dutch colony. In 1907 this company became known as the Bataafsche Petroleum Maatschappij (BPM, Batavian Petroleum Company). Today Royal Dutch Shell (RDS) is also known as the Royal Dutch, or even more commonly as just Shell, due to the amalgamation of the BPM with the British Shell Transport and Trading Company from 1907 onwards.

Shell explored and transported oil, first from the Dutch East Indies and later also from other parts of the world; they did some refining and sold products such as petroleum and paraffin, and somewhat later gasoline ('benzine' in Dutch).

In the early years of oil and gas production, up until about 1930, empiricism ruled. The first laboratory, opened in Amsterdam in 1913–1914, was mainly used to test the quality of the various products and to do chemical research. In 'proeffabrieken' (pilot plants) production and refinery processes were tested. Flow issues hardly got any attention.

But there was one flow problem which did get attention from the oil companies almost from the start of the 20th century. Forbes and O'Beirne (1957): "During the second half of the 1900–1927 period the engineers also acquired reliable formulae on the pressure needed to overcome the friction in pipelines. The existing theories of the flow of liquids (mainly water) were completed and adjusted for practical application in the petroleum industry, largely by practical tests with existing pipelines and others built for experimental purposes. Those of the first category were conducted at first by the oil companies themselves; contact between the companies was slight and each one developed

↑From the street the Laboratorium voor Fysische Technologie in Delft (later known as Kramers Laboratorium) still looks pretty much the same as it did after its opening in 1949. A plate at the main entrance still shows its relationship with Shell. The impressive tower was partly meant to house a high distillation installation (pilot plant) but this was never realized. Part of the tower housed an installation where students could learn to calibrate measuring flanges; this had a water tank at 19 m height. The white building was built in the 1960s and become known as The White Elephant.

and used its own formulae. The basis for reliable calculations of the pipelines of RDS was laid by experiments made by J. H. C. de Brey at Batavia [now Jakarta, Indonesia] in 1913. Without reliable data on friction it is impossible to plan the most economic design for a pipeline. This includes the pipe diameter, the pressure to be applied and hence the strength of the material used for the tubing, the number of pumping stations required to make good the pressure lost by friction, etc."

In the 1920s the Amsterdam laboratory of Shell rapidly expanded and research became more professional and extensive. For many years it would be the largest industrial laboratory in The Netherlands. No longer was the research directly linked to problems which arose at the plants. More 'exploratory' research was initiated, especially on chemicals. Around 1930, the Wetenschappelijke Afdeling (Scientific Department) started research on various topics. One of them was the behaviour of drilling fluids, a theme which is related to the field of colloid chemistry. As the boreholes were drilled to larger depths over the years, knowing the viscous behaviour of the fluids became even more important.

In 1926 the Central Production Department of Shell was set up in The Hague and the following year the mining engineer and hydrologist Jan Versluys (1880–1935) started to work there. His publications in the Proceedings of the KNAW (the Academy of Sciences) reflect the broad range of topics that he tackled. Besides physical studies into the occurrence (and flow) of oil, gas, and water in reservoir rocks, Versluys studied the vertical extraction of oil and gas by natural flow

and gas lift. He succeeded in giving a scientific explanation of surges in flowing wells from which he could give recommendations on how to make the flows more regular. Another example of his work: predicting the correct aperture of sand screens, used downhole at the inlet to the well tubing pipes in relation to the size range of the sand grains present. Production techniques themselves made rapid progress. In the 1930s Versluys and other investigators had written mathematical treatises on two-phase vertical flow in well tubes, both with and without gas lift. Results obtained with different diameters of tubing were recorded, while a correlation for flow and air-lift from the annulus space between tubing and casing of wells was also established. Versluys' mathematical treatment of the production engineering field was adopted by various American investigators.

From 1928 a second laboratory was used by Shell for more scientific research: the Proefstation Delft. Here the topics were related to the combustion engine, which had become very common during the 1920s. The diesel engine was up and coming, and diesel fuel became one of the fuels which were investigated in Delft. Lubricants for engines also got the attention of the (small group of) researchers there. After World War II the engine research largely moved to England but some was still done in Amsterdam. The laboratory in Delft focused on improvement for the industrial processes and equipment which were used by Shell in its refineries and chemical plants.

The Burgers Archives in Delft contain no evidence that Burgers ever did any consultancy work for Shell. He was however

## J.O. HINZE (1907–1993)

Oscar Hinze was a student of Mechanical Engineering in Delft in the early 1930s and became attracted, as one of the few at that time, to fluid mechanics. The fact that Burgers' laboratory was not integrated in the main building of Mechanical Engineering may have contributed to the unfamiliarity of the field to most people at the department. Hinze appeared to be a clever student and later assistant (one of the courses he followed in Delft was 'Relativistic mechanics' in 1934). In his inaugural address he remembers that Burgers gave him the task of making a movie of some flow phenomena, much to his surprise.

Maybe it was the lack of good prospects at the TH for a young engineer in the middle of the 1930s, with the financial crisis still lingering, but for some reason Hinze decided to work for Shell in their Proefstation in Delft in 1935. There, in 1952, he became head of a new department concerned with engineering research on manufacturing. In this position he continued his studies on the fundamental as well as the technical aspects of turbulence in relation to flow resistance, heat and mass transfer, and mixing.

By 1955 Burgers had left Delft but he hadn't left his Laboratory professor-less since Broer had already been appointed in 1949 and taken over supervision during the 1950s, together with lecturer (lector) Leo van de Putte. In 1955 the future looked bright for the TH and for the department of Mechanical Engineering. Soon new buildings would be opened in the Wippolder, to the south of Delft, and the financial future also looked rosy. Under these circumstances plans were made to attract a second professor in fluid mechanics. Since Broer had been trained as a theoretical physicist, it seemed only natural to find someone with a background in the engineering world for this post. Thus Hinze returned to the TH in July 1956, as professor of fluid mechanics, but he would remain advisor to Shell. In 1977 he retired and left fluid mechanics completely.

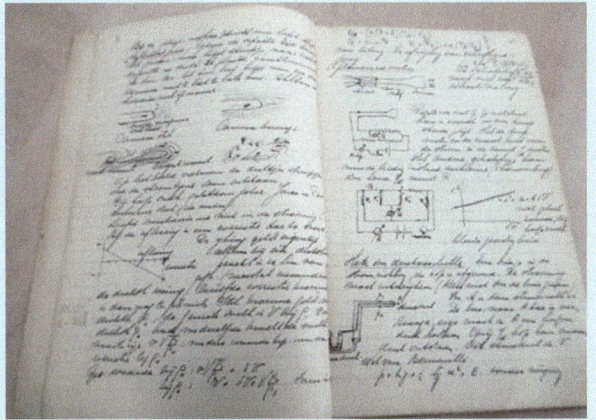

I↑n the Burgers Archives in Delft one can still find some hand-written lecture notes from the 1930s which probably were produced by Hinze. This is the 'repetitorium' of the 'Introductory lectures' on fluid mechanics given by Burgers around 1935. (courtesy of Burgers Archives / TU Delft)

←Hinze has become well known for his book "Turbulence" whose first edition was published (by McGraw-Hill of New York) in 1959, contained 586 pages and had as subtitle: "an introduction to its mechanism and theory". The second edition of 1975 had 790 pages but had lost the subtitle. In fact, the primary version of this book had appeared as a two volume 'course on turbulence' for Shell personnel in 1951 after a series of lectures by Hinze. In the preface to this course Hinze referred to an earlier report by Van der Hegge Zijnen, entitled "Turbulence – ally and enemy", which had appeared in 1948. About the strong theoretical nature of his course, Hinze explained that he had been asked to take this approach "since it was felt that many problems in chemical engineering can only be solved with success if treated as fundamentally as possible". This is the copy that hydraulics professor Jurjen Battjes bought in the USA in June 1966 for US$17.

contacted by the BPM in 1937 regarding the extreme nuisance caused by strong winds in one of their offices on the island of Curaçao. The Delft Laboratory did some nice experiments on this.
Indirectly Burgers became important for Shell when one of his few students during the 1930s started to work at Shell in 1935: Oscar Hinze. After Hinze received his master's degree in 1933 he was an assistant at Burgers' laboratory until he left for the Proefstation in Delft. There he became involved in research related to fluid dynamical and heat trans-

fer problems. It is plausible that it was Hinze who would arrange a position for Burgers' most prominent scientific assistant since 1921, Van der Hegge Zijnen. It seems that tensions between Van der Hegge and Burgers had grown severely during the War and in 1946 Van der Hegge took the opportunity to start working for Shell in Delft. There he had the privilege of having his own wind tunnel, the measuring section of which crossed his office room! There he retired in 1957 and he died in 1968, 71 years old and lonely.

Chemical engineering, as it was called, started to flourish from the middle of the 1930s. In 1939 the Amsterdam laboratory was expanding again, housing more than 1200 staff. Besides chemical processes like distillation and extraction, physical processes like stirring and mixing got more and more attention. From 1949 the laboratory was known as KSLA (Koninklijke/Shell Laboratorium Amsterdam). New, more fundamental research groups were formed, and new areas of research were explored, e.g., the rheology of polymers. The number of engineers, chemists, and physicists increased strongly, and they got more support from mathematicians (in 1955 Shell was the first Dutch company with a computer for scientific work). The foundation of the department called Process Engineering in 1956 was an important stimulant for new research in fluid mechanics. The opening of a special two-phase flow installation at KSLA in 1959 marked the growing interest in these kinds of flows. But it was only in 1968 that researchers at KSLA could use a wind tunnel. One of the first projects for which it was used was the flow around the very high chimney which was under construction at the Shell refinery in Pernis, near Rotterdam.

In the immediate post-war years Shell evaluated and reorganized its research activities. The number of patented inventions started to increase considerably. Delft (which would move to Rijswijk in 1960) and Amsterdam remained important research facilities. In 1946 the BPM decided to offer three million guilders to the Dutch State to invest in three academic institutions. One of them still had to be built on the new 'campus' of the TH Delft: the Proeffabriek voor Physische Technologie (Pilot Plant for Physical Technology). The BPM also supported the new plant by sharing the experience of its scientific and technical staff. It had already done this before the war with the appointment in 1936 of the 'special professor' in Physical Technology, Willem van Dijck (1899–1969), who had been the first physicist of BPM in 1927.
It was Van Dijck who convinced Shell that Delft needed a laboratory for physical technology, besides the already existing laboratory for chemical technology. Van Dijck was succeeded in 1947 by a young BPM physical engineer named Hans Kramers (see § 5.1.4), who asked Shell to change its plans for Delft: a pilot plant would not be very useful for physics students who had to be trained in knowledge of several basic processes, not in making particular products. The Laboratorium voor Fysische Technologie was officially opened in 1949 and transferred to the TH Delft in 1951.

# 3.4 FLUID MECHANICS IN DAILY LIFE

## • 3.4.1 TRANSFORMING THE COUNTRY: FROM ZUIDERZEE TO IJSSELMEER

For ages large parts of The Netherlands felt the influence of the salty Zuiderzee (Southern Sea), a shallow bay of the North Sea in the northwest of the country. During the 19th century plans for the impoldering of the Zuiderzee became more concrete; the extension of the fertile area for the growing population being one of the arguments. The decisive trigger to start the closing ('restraint' may be a better word here) of the Zuiderzee came in 1916. A heavy storm surge had caused severe floods at several places around the Zuiderzee, including parts of Amsterdam. To many it was clear that the Zuiderzee had to be transformed from a sea into a lake by means of a long dike in the water, a dike which would become known as the Afsluitdijk. To this end, the Zuiderzeewet (Southern Sea Law) was passed in 1918.
The minister responsible for the waterways at that time was Cornelis Lely (it was the third time he had been a minister; see also § 2.1.1) and he decided to set up an independent study group for this huge project, de Dienst der Zuiderzeewerken (DZ). The RWS was not involved and most of their engineers didn't mind this at all: they were quite sceptical about the feasibility of the project.
Lely had already worked on a plan for the impoldering in the late 19th century but realized that it would be necessary to have some experts involved in the project. The already famous physicist Hendrik Lorentz, recently retired, was asked to become chairman of the Staatscommissie (state committee) for the Zuiderzeewerken. One of the main tasks of the committee would be the question: what influence would a causeway (between the northern part of the province of Noord-Holland and the province of Friesland) have on the water levels, especially during high tide, near the coasts of the North Sea and the Wadden Sea? Would it be necessary to strengthen or enlarge the dikes?

In 1918 hydraulic engineering was still mainly a field in which experience was leading, and the formulas and methods used in hydraulics were mainly based on empirical laws and rules. For the complicated task which the Committee had to tackle, a more fundamental, theoretical approach was necessary, based on the laws of hydrodynamics.
In 1927, when the committee had finished its work, Lorentz confessed that he had been apprehensive about the task he had been given, as: "a physicist is not used to working on problems of such high complexity and with so few well-determined data". The committee had started with collecting data of tidal movements, flow patterns, and the effect of

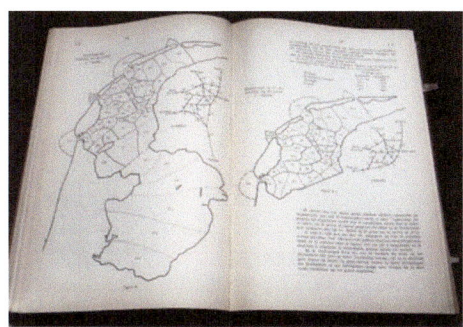

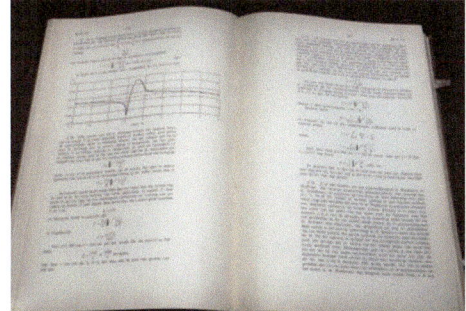

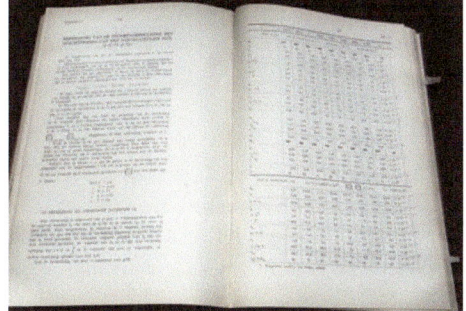

↑In September 1926 the State Committee led by the old professor Lorentz published its final report. Its 350 pages also contained many maps of the Zuiderzee area, tables of water levels, and a concise explanation of the calculations which had been performed. In the list of references one finds the famous textbook Hydrodynamics by Horace Lamb. (from: Staatscommissie Lorentz (1926))

↑On the occasion of the opening of the Afsluitdijk to the public in 1933 (the last gaps in the dike were closed in 1932), a lot of memorabilia were put on the market (this booklet was published by Wilms of Wieringen). This is a set of postcards, showing the completion of the dike and the currents which the builders had to face. These currents became stronger as the gaps became narrower.

←The activities around the Zuiderzee attracted national interest during the 1920s. Dutch citizens could buy special postcards showing the land reclamation (landwinning) plans, the origins of which came from Cornelis Lely himself. The large areas of new land had to be 'won' from the sea: the Noordoostpolder, Flevoland, and the Markerwaard. The last one, the western area, has not yet been realized.

↑The Zuiderzeewerken also meant the appearance of new kinds of buildings in the Dutch landscape. This is the Lely pumping station near Medemblik, which was meant to drain the Wieringermeer (a new polder just south of the western end of the Afsluitdijk). It is still functioning today and as beautiful as it was when opened in 1930. It was designed by the architect Dirk Roosenburg (grandfather of the well-known architect Rem Koolhaas).

the wind on the water. Then, two methods that had been proposed by others for the calculations on the effects of this closing were investigated but rejected. Finally, Lorentz developed his own method, based on equations from fluid mechanics and a linearization of the flow resistance. Since the water in the area followed an extensive network of gullies, it appeared to be possible to model the Zuiderzee as a system of canals. One of the decisions which was made on the basis of the results found by the Lorentz Committee was to make a dogleg in de Afsluitdijk so that it would land in Friesland a few kilometers further to the north.

The Committee was not sure what would happen during storm surges after the Afsluitdijk had been built. This was partly due to the fact that flow rates in the access points to the open sea (zeegaten) were unknown. Furthermore, no measurements had even been done during storms and no one knew exactly what the effect of the Earth's rotation would be. Lorentz et al., decided to add a safety margin of 20%. Reading the impressive final report of the Committee, one can only admire the way in which Lorentz had been able to simplify the formulas from fluid mechanics without destroying their 'character' and making the results nonsensical. The very work-intensive, and rather monotonous, calculations were all made by a few people from the RWS with help of only a handful of 19th century mechanical calculators. Besides calculations, laboratory experiments were done to find answers. Around 1920 one of the engineers working for the DZ had constructed a wooden basin of 15m length in the garden of the DZ office. Primitive though it was, it contained a construction to generate waves and at the other end of the basin several types of dikes with variable slopes could be tested. The first wind tunnel at the RSL (see § 3.2.1) was also used to do some Zuiderzee research. But the more serious experiments could still not be performed in The Netherlands. For example, for the investigations into the behaviour of floodgates, the DZ used the possibilities offered by the model basin in Karlsruhe (see also § 3.2.2), despite the suspicion against this 'scientific' method among many engineers in the field of hydraulics. The DZ staff had to regularly defend their approach.

↑In 1926 Von Baumhauer did some experiments with model sails in the wind tunnel of the RSL. The four sails (typical for all old Dutch windmills) drove a dynamo with which the power output could be determined. Von Baumhauer soon concluded that with just some adaptation of the old sails no important improvements in efficiency could be achieved. Nevertheless, the RSL continued to perform measurements on windmill models. (courtesy of Stichting Behoud Erfgoed NLR)

Presumably, it was Jo Thijsse, director of the Waterloop-kundig Laboratorium (see § 3.2.2) and an important man in the DZ, who pulled Burgers into the Zuiderzee Project from about 1928. Burgers' main contribution was related to the pumping-engines, situated around the Zuiderzee. Together with Van der Hegge Zijnen, Burgers performed extensive and time-consuming calculations on the flow in the fan of a centrifugal pump for the Lely pumping station at Wieringer-meer (see § 6.3 for more details).

After the completion of the Afsluitdijk, the impoldering of the IJsselmeer, the new name of the Zuiderzee, was continued. In 1936 work on the Noordoostpolder was started but it was only completely drained in 1942. After the war the draining of Flevoland followed, leading to Oostelijk Flevoland in 1957 and Zuidelijk Flevoland in 1968.

In the meantime the attention of most hydraulic engineers had been directed to another part of the country. Even In the 1930s, the young RWS engineer Johan van Veen (1893–1959)

had warned that the dikes in Zeeland, the south-west corner of The Netherlands, should be heightened without delay, since the danger of a disaster due to a storm surge was becoming quite real. In the province of Zeeland, and part of Zuid-Holland, the sea arms of the Delta area were still open and the tides influenced a large part of the waterways. Coastal protection had no priority after the war. In 1953 Van Veen's fear became reality: several parts of the area were flooded and more than 1800 people were killed.

The 'Flood Disaster' led to a general consensus that the sea arms had to be closed. The Delta Plan, which Van Veen had already formulated long before 1953 and parts of which had indeed been realized, would be the main project of RWS for the next decades. One of the consequences of the intensified activities would be the construction of one of the first (analogue) computers in the country, the DELTAR (see § 6.3).

### • 3.4.2 WINDMILLS

For centuries the Dutch had been surrounded by windmills. Several types had been developed and though there must have been some innovation from time to time, the windmills of about 1900 seem to have been hardly more efficient than those of 1700.

In the Netherlands, the so-called common sail had been predominant since the Middle Ages. During the nineteenth and twentieth centuries, some Dutch millwrights developed new windmill sails. Their aim was to make them more efficient aerodynamically and to make the mill operation easier. This was an effort to keep the traditional windmills economically viable in competition with factories running on steam or combustion engines and with electric pumping stations.

In the late 1920s a strong increase in the number of inventions regarding sails can be seen in The Netherlands. A millwright named Dekker took out a patent in which the four stocks of the mill were given an aerofoil shape on the side which 'faced' the air by completely covering them with galvanised steel plates. 'Dekkerised sails' provided enough surface area to be able to work the mill with no sailcloth spread if the wind was strong enough.

Dekker's work incited others to improve on his sails or to develop other innovations. An engineer named Fauël was inspired by the jib ('fok' in Dutch) on a sailing boat. In his design the leading boards of the windmill's sail were replaced by a rounded profile of wooden slats in the form of a foresail, leaving a small slot between this profile and the stock. Its working principle can be compared to a leading-edge slot on an aircraft wing. It enabled the mill to work at a lower windspeed, but in variable windspeeds tended to make running the mill at a steady pace difficult. For this reason it was often equipped with air brakes operating by centrifugal force. Today, the jib-sail and improved versions of the Dekker sail are still used on Dutch windmills.

↑In 1939 new sails were mounted on the Prinsenmolen near Rotterdam. The experiments showed that, when the wind came from the North or the West, the power output of the mill was indeed higher than with the traditional sails. (from: Havinga (1959))

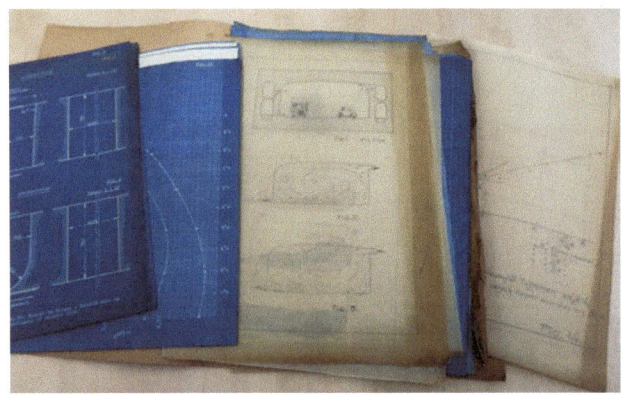

↑Documents from the Velsertunnel files in the Burgers Archives in Delft. Burgers and his staff did experiments and calculations on this tunnel under the North Sea Canal near Beverwijk in the late 1930s, and in 1941 the construction of the tunnel was started. After the war the design was changed and the tunnel was only opened in 1957. (courtesy of Burgers Archives / TU Delft)

↑To get a good three-dimensional picture of the rather complicated system of the ventilation channels of the Maastunnel, a wooden model was built in Delft. This picture was probably taken in the temporary wooden 'shed' that was erected near the Laboratory for the model experiments. (courtesy of Burgers Archives, TU Delft)

←The characteristic ventilation buildings of the Maastunnel, 37 meters above ground level and 26 meters below, have been part of the daily life of the Rotterdammers for more than 75 years. They were designed by the architect Ad van der Steur, who would also be responsible for the new building of the faculty of Mechanical Engineering of the TH Delft in the 1950s. Today the Maastunnel is a national monument and the pride of many in the city. The fans of the ventilation system are hardly rotating anymore: the CO levels in the tunnel don't reach critical values anymore since cars have become much cleaner. (Wikimedia Commons / photo by S.J. de Waard / CC BY 3.0)

One of the first to show scientific interest in improving traditional windmills was Albert von Baumhauer (see § 7.1 for more about him). He was probably influenced by the words of Van Iterson written in 1911 (see the introduction of this chapter). In 1918 Von Baumhauer, who was then working for a Dutch aircraft company, had already designed sails which in fact were propeller blades with cross sections similar to wing profiles. After moving to the RSL in 1921, he continued his research both on an existing old windmill and on models in the RSL wind tunnel. In 1923 he got involved in the newly-founded Society for the Preservation of Mills in The Netherlands "De Hollandsche Molen" (which still exists today). Although by 1930 windmills had generally been replaced by engine-driven pumping-engines, some windmills were still considered to have a useful function for drainage. One of them was the Prinsenmolen near Rotterdam, originally built in 1648. In 1935, its administrator asked Burgers and his colleague Muysken to study the mill's vanes: could the shape of the vanes be changed, to improve the working of the mill and to gain experience which could be useful to other mills involved in drainage? Subsequently, the Royal Netherlands Institute of Engineers (KIvI) approved sponsoring the work and set up the Prinsenmolen Committee of which Burgers and Muysken became members, and of which Adolf Havinga, a former student of Burgers, became the very active secretary.

From 1936 Burgers and his staff got involved in theoretical and experimental research on windmill sails. A scale model of the 'tower' of the Prinsenmolen was put in one of the Delft wind tunnels to investigate the wind patterns around the mill. Other experiments were done in a meadow outside Delft on a rather small but fully operational windmill. The research done on the Delft windmill led to a 'conformity rule' with which it was possible to predict the performance of a full-scale mill from measurements done on a model. The members of the Committee claimed that for the very first time they had been able to include unsteady air flow (i.e., a more realistic representation of wind) in their experiments and theory.

In 1939 the real mill was tested with newly developed sails and it appeared that its efficiency had indeed increased. The experiments had also taught the researchers that one of the most difficult questions to answer was: how do you measure the wind speed around a windmill? At what height(s) do you have to measure, and which figure can represent the fluctuating data that you get from the anemometer?

## • 3.4.3 VENTILATION OF TUNNELS

The Netherlands had always been a land of bridges, and the city of Rotterdam had been no exception in this respect. But when in the late 1920s the city council started to think (again) about a new connection between the north and south shores of the river Maas, it started to consider the building of a tunnel. Traffic tunnels under rivers were still very rare in the world at that time.

Due to this interest by the Rotterdam council, the engineering world in The Netherlands started to study the technical questions related to traffic tunnels. In 1933 the council had decided, in principle, to go for a tunnel and soon after at the Royal Netherlands Society of Engineers (KIvI, founded in 1847) a commission for 'tunnel ventilation' became active. Though the final decision to build a four-lane tunnel for motor vehicles plus separate lanes for bicycles was only taken in January 1937, the Werkspoor company (see § 3.3.1) was already busy studying the ventilation issue. Burgers was asked by chief engineer Theunissen (with whom he was already familiar thanks to his advice on the ventilation of the State Mines and their personal meetings) to design the ventilation system and to test a model in his Laboratory. This work started in the Delft Laboratory in 1934 and would last for nearly five years. During this time, Burgers published a paper on the problems related to traffic tunnels, which he and Van der Hegge Zijnen had found in the literature and during visits to foreign tunnels.

Burgers and his team did model experiments for about two years, not only on the ventilation channels but also on the flows within the tunnels. They also investigated the influence of the wind flowing around the ventilation buildings on their performance: a model of one of the buildings was put near a fan and smoke was blown through the channels inside. In the spring of 1938 the Delft team could finally write down which characteristics they thought the fans should have. By the summer of that year they had produced 31 reports. In early 1941 the tunnel construction was finished while traffic was already passing through the Maastunnel. Due to the ongoing war, no official opening took place but from February 1942 the tunnel was fully operational. From March that same year Burgers and his staff regularly visited Rotterdam to test the ventilation system. Later that year they finished their 38th report.

In 1942 the Maastunnel was the biggest traffic tunnel in the world. In a letter, Burgers thanked the chief engineer of the tunnel project for the: "friendship and respect which one has met with each other … To me it is a great honour to have been part of this work. The memory of it I shall always keep in mind, as well as the memory of the people with whom I have made acquaintance".

His work for the Maastunnel made Burgers a respected expert in tunnel ventilation. Besides that, he had definitely established his fame as an 'engineer' within the engineering community. From the late 1930s till the 1950s Burgers also gave advice on other tunnels for which plans were being made: the Velsertunnel near Amsterdam (model experiments in 1937–1939), the IJ-tunnel in Amsterdam (1939, opening in 1968!), and even on some tunnels in Tunisia.

© Springer Nature Switzerland AG 2019
F. Alkemade, *A Century of Fluid Mechanics in The Netherlands*, https://doi.org/10.1007/978-3-030-03586-0_4

# 04

## DEVELOPMENTS AFTER 1955

Up until the Second World War fluid mechanics remained the field of interest of a relatively small group of people, scientists and engineers, at a limited number of academic groups, institutes, and industrial companies. In the immediate post-war period the money and knowledge needed to get fluid mechanics back on the rails were still lacking. But as reconstruction progressed industry and the economy started to recover, beginning to flourish in the late 1950s. At the same time fluid mechanics started to flower and expand.

53

The only two professors of fluid mechanics in The Netherlands around 1950 in front of their new housing in Delft, close to the old one of 1921. Burgers is the fourth from the left and Broer is the second from the right. The man with the pipe in his hand is Leo van der Putte, head of the laboratory and reader in fluid mechanics for many years. Also on the photo are Broer's daughters; one of them has become well-known as the poet and novelist Anna Enquist. (courtesy of Burgers Archives / TU Delft)

# 4.1 UNIVERSITIES

By 1955 The Netherlands had two professors for fluid mechanics: Jan Burgers and Bert Broer (1916–1991). Both were trained as physicists and were able to apply mathematical techniques which were (and maybe still are) regular in physics but still rather unknown to the engineering world of the 1950s. The question then arises whether the Delft 'school' had caused an increase in the interest in fluid mechanics among the physicists at the universities in The Netherlands. In fact, the foundation of two new technical universities led to a growth in the number of research activities in this field, and not only in the 'traditional' circles of mechanical engineering.

### • 4.1.1 DELFT

In 1949 Burgers' second laboratory – for which he had pleaded for about two decades - had finally been realized. But it would take another ten years before a 'definite' Laboratory for Aero- and Hydrodynamics (AHD, also known as the Vakgroep Stromingsleer from the early 1970s) would be built on the new TH campus on the outskirts of the city. At the department of Mechanical Engineering an engineer, Hinze, had been appointed in 1956 as successor to Burgers (see § 3.3). In his inaugural address he stressed the fact that his field would be 'technical fluid mechanics'. At Shell Hinze had noticed that empirical formulas were usually preferred (as an anecdote he told his audience that in 1939 engineers, presumably at Shell, still used a handbook containing about sixty of these formulas!) and he saw it as one of his tasks

to provide those working in the field of industrial flows with formulas derived from fundamental theory. From the time of Hinze's appointment, one could say, fluid mechanics in The Netherlands became more and more under the influence of the industry (especially Shell). In the 1990s, when Shell and other companies started to diminish their efforts in fundamental research, this influence would also decline.

In retrospect one gets the impression that Burgers had not really been able to develop a legacy of staff members and students to follow him in Delft, and he realized that himself when he left in 1955. The study of fluid mechanics in Delft had not changed since before the War, but after 1945, one could say, Delft started to lose its world-wide renowned position in fluid mechanics (but not in hydraulics!). In the 1950s the number of students at the Laboratory of Aero- and Hydrodynamics was still small, and most of them seem to have been mainly active in experimental work (for which Ritzo de Haan, who came from the NLL, was the supervisor). In 1959 five PhD students were working in the lab and about sixteen 'ordinary' students (half of them studying Mechanical Engineering, the other half Technical Physics). Though Broer had been trained as a theoretical physicist, he was also involved in these experiments. The main topics at AHD at that time were compressible flows (supersonic jet, shock waves) and rheology; also something quite new at the time: magnetohydrodynamics. From the late 1950s, when Hinze had become professor, turbulence would get more attention and remains a main topic today. It was probably also due to his influence that some experiments were started

←In the 1950s TH Delft found a new habitat in the polder to the south of the city. Here would be the new building of the Department of Mechanical Engineering, but like in the pre-war situation the new Laboratory for Aerodynamics and Hydrodynamics would not be part of the 'mother' premises. The new AHD building was erected to the west, along the old road to Rotterdam. It was only completed in 1961, though the drawings had been made (by the famous architects Van den Broek and Bakema) in 1957. The tower was specially built for experiments on two-phase flows in long tubes but has not been used very intensively. The building in the foreground housed the high-speed wind tunnel; it had a special construction: in case of an explosion only the roof would come off. (courtesy of Burgers Archives / TU Delft / photo by Fons Alkemade)

on two-phase flows: in a closed-water circuit, suspensions of granular material could be investigated and a 'fluid-air ejector' was built in 1959. Cooperation with other academic groups, with institutes, and with industry seem to have hardly existed around 1960.

When Broer left Delft in 1962 and went to the still young TH in Eindhoven, Henk Merk (1920–1995) was appointed for both Mechanical Engineering and Technical Physics. Merk had been trained as a physical engineer in Delft and, like Hinze, had worked at the Shell laboratory in Delft. Later he had worked at Shell's KSLA in Amsterdam. Merk's supervisor during his PhD research had been professor Jan Prins (1899–1986), a physicist who had become interested in heat transfer when a professor of physics and meteorology at the Agricultural University in Wageningen. In Delft he would be the first professor, from 1946, in the field of (fundamental) heat transfer. Merk's thesis was entitled Stofoverdracht in laminaire grenslagen door gedwongen convectie (Transfer of matter in laminar boundary layers due to forced convection). At the TH Merk would become mainly concerned with rheology, a still rather unknown topic at that time. The lecture notes for the courses in rheology which he wrote with Gerard Kuiken in the 1970s were also used in Eindhoven and Twente. (Gerard Kuiken's brother Henk would publish papers on fluid mechanics when working at Philips.)

Though TH Delft had already had Technical Physics students since 1928 and an independent Department of Technical Physics since 1948, the very first ordinary professor appoint-

ed in the field of fluid mechanics (and also of heat transfer) was Charles Hoogendoorn (1930–2012). He had been a student of Burgers', came from Shell and started his job in Delft in... 1970. This was ten years after professor Broer had stressed the importance of good contacts with Technical Physics in Delft during his lecture on fluid mechanics on the occasion of the 188th dies natalis of the Delft University in 1960 (in which he had talked about flames, magnetohydrodynamics, plasmas, and nuclear fission).

Hinze had done some original work at Shell (on droplets) but in Delft his research activities were limited, and he didn't have many PhD students. One of them was Allan Chesters whose thesis was entitled Fundamental problems in gas-liquid two-phase flow (1978, Hinze supervisor). Hinze was succeeded in 1978 by Gijs Ooms (1941) who had also worked at Shell. In 1986 Frans Nieuwstadt (1946–2005) left the KNMI and became for many years one of the best-known representatives of fluid mechanics in Delft. He had finished his PhD (on the nocturnal boundary layer) in 1981 at the Vrije Universiteit in Amsterdam under the guidance of Tennekes (see Ch 7), and Steketee (see below). During his professorship research on turbulent flows, especially the role of coherent structures and drag reduction, would get much attention.

The Department of Mechanical Engineering had some other sections where research related to fluid mechanics was done. The Laboratorium voor Warmtetechniek (Laboratory for Heat Technology) had been in existence from 1911 and later became known as the Laboratorium voor Warmte- en

↑In 1967 Aeronautical Engineering finally got its own building, on the outskirts of the Delft campus. There was plenty of space to build another aerodynamical laboratory there, and it was opened in 1969 and can be seen on the left. The main building (on the right) is 13 stories high and can be seen as a firm statement of the Department which was still not really independent even in 1967. The high-speed wind tunnel in the building in the foreground was built by Werkspoor and the first measurements were performed in 1972 after a long period of calibration. In 1973 the first supersonic wind tunnel of the NLR/NLL, also the first in The Netherlands, was donated to the Delft laboratory and was deployed for students' practicum. For forty years Peter Bakker (1945) worked among the wind tunnels, from 1989 as professor of aerodynamics.

←Towing tanks always fascinate a lot of people. In 1955 TH Delft celebrated its fiftieth anniversary as Technical University and invited all the people from Delft, and elsewhere, to come and see what was going on in some of the laboratories. The new Shipbuilding Laboratory had just been opened and demonstrations were given about the towing of model ships and what measurements were done. (courtesy of TU Delft / photo by Fotografische Dienst via Beeldbank / CC BY)

Stoftechniek (Laboratory for Heat and Dust Technology). It got a new building in 1956, next to that of AHD. Still later this group became known as the Laboratory for Thermal Power Engineering and in the 1980s some research on two-phase flow was carried out there. In the 1980s the research group in this area would become known as Process and Energy. Hydrodynamical problems related to lubrication were studied at the section for Tribology (which had its origins in the 1950s, and also had a professor who came from Shell). A rather remarkable building close to that of AHD was erected in 1977 to become known as API: Laboratorium Apparatenbouw Procesindustrie (Equipment Building for the Process Industry). Though mainly intended for engineering research, flow phenomena had to also sometimes be considered. Last, but not least, the Dredging Laboratory must be mentioned (see § 6.2), where from the 1980s research has been performed for a branch where the scientific approach had long been absent (see also § 6.2).

Burgers had already given lectures on the aerodynamics of airplanes during the 1930s, but students could not graduate as an aeronautical engineer. Only in 1940 did Aeronautical Engineering get an official status, i.e., became a 'sub-department' of Mechanical Engineering. It took another 35 years before it would become an independent department! The first professor was Henk van der Maas (1899–1987) who would remain the face of 'Vliegtuigbouwkunde' (later: 'Luchtvaart- en Ruimtevaarttechniek', translated as Aerospace Engineering) till 1967. But he would also play important roles in several other organisations related to aeronautics; for years, e.g., he was chairman of the board of the NLR/NLL.
Van der Maas had been trained as a student in Ship Engineering in Delft. Burgers had been the supervisor of his PhD thesis in 1929 but Van der Maas has never really been involved in aerodynamics. He worked at the RSL/NLL, mainly as a test pilot for research purposes. Though in 1953 the first wind tunnel for aeronautical purposes was finished (see § 6.2.1), there was still no professor of aerodynamics employed by the sub-department. Lectures were given by Burgers and Broer. Only in 1960 the first 'aeronautical aerodynamics' professor was appointed, Jaap Steketee (1927–1998). Steketee had been an assistant in Burgers' laboratory from 1948 to 1950 and had done his PhD at the university of Toronto, Canada. Less than two years later he was joined by an extraordinary professor for 'supersonic aerodynamics', Siegfried Erdmann (1916–2002). During the War Erdmann had worked on rockets in the well-known Aerodynamics Institute in Peenemünde, under the guidance of Wernher von Braun. In 1946 he was invited by the NLL/NLR to help in building their first high-speed wind tunnel (see also § 6.2.1). Jan van Ingen (1932) published his PhD thesis on 'incompressible laminar boundary layers with and without suction'

in 1965 under the guidance of Steketee and Van der Maas. In 1970 he also became professor of aerodynamics. During the 1980s the Aerodynamics section worked on several themes, e.g., gas dynamics, MHD (in the context of the re-entry of spacecrafts), transonic and supersonic flows, flow around wing profiles, and flows involving chemical reactions (in the context of rocket propulsion).

Before 1940 fluid mechanics got only minor attention from Scheepsbouwkunde (Naval Architecture or Naval Engineering), which was then a subdivision of Mechanical Engineering. However, already in 1906 discussion arose about the question of whether a model basin should be built in Delft. But since the naval industry showed only small willingness to sponsor the expensive installation, nothing came of it. It was only in 1937 that the first model basin could be put into operation. When the new building of Mechanical Engineering and Naval Architecture was opened in 1956, also a new (and larger) model basin could be used. At the end of the 20th century the section of Ship Hydrodynamics had two basins (one deep, one shallow) and a small cavitation tunnel. The first professor in Delft who was able to do serious research on ship hydrodynamics was Van Lammeren of the NSP (see § 3.2.3) who was extraordinary professor in Delft from 1952 till 1962. He was succeeded by another NSP director, Dick van Manen (1923–2006). Van Manen had published his PhD thesis on the design of ship's propellers in 1951 with Troost (see also § 3.2.3) as supervisor. Only in 1961 was the first ordinary professor for ship hydrodynamics appointed in Delft, Jelle Gerritsma (1924). In 1955 he had become head of the Shipbuilding Laboratory, as it was called, and from 1958 he had lectured on ship movements. The main topics of the Laboratory were (and still are) the resistance on and propulsion of vessels and their dynamical behaviour. Delft became especially active in research on fast and advanced ship models.

Thijsse (see § 3.2) became extraordinary professor of Hydraulics at the department of Weg- en Waterbouwkunde (later called Civil Engineering) in Delft in 1938, and from 1946 this became an ordinary professorship. After the Storm Surge of 1953 Thijsse got involved in the huge projects related to the Deltawerken (Delta Works) which had to protect the southwest of The Netherlands from the sea. Not surprisingly, his research group at the TH also got involved and remained so for many years. Still in 1985, for example, experiments and numerical experiments were done for RWS related to secondary flows in the Oosterschelde (where in 1986 the famous Storm Surge Barrier was completed). During the 1950s Dutch hydraulic engineers increasingly realized that their knowledge would be useful for many other countries. In 1954 Louis Mostertman (1920–2007) became employed in Thijsse's group and was soon asked to start preparations for an International Course in Hydraulics which would be given in Delft to students and engineers from developing countries. Mostertman became director of what became known as the IHE: the International Institute for Hydraulic and Environmental Engineering. In 1962 he became professor in Delft.

When Thijsse retired in 1963, he was succeeded by Willem Bisschoff van Heemskerck (1921–1973), who had worked at RWS. In 1966 Johan Schönfeld (1918–2005) became lector (associate professor). He had been trained as a physicist in Delft and had written a PhD thesis on tides and waves (1951) when already employed at RWS. Schönfeld is considered as a pioneer in computational hydraulics in The Netherlands, especially for shallow waters. In 1968 Eco Bijker (1924–2012) left the WL laboratory De Voorst, where he had been director, and became professor in Delft, specializing in coastal engineering.

Around 1970 the new Stevin III Laboratory, better known as the Laboratorium voor Vloeistofmechanics (Laboratory for Liquid Mechanics) was opened as a new part of the building for Civil Engineering. It had a huge experimental hall which was 8 meters high. In the same period the Delft hydraulics group got involved in an ambitious project initiated by RWS and a foundation which included the WL: Fundamenteel Onderzoek Waterstaat (Fundamental Research Public Works Ministry). Topics included in this project were transport of sediments and 'coastal research'. Some years later the name of the project was changed to Toegepast Onderzoek Waterstaat, TOW (Applied Research PW Ministry) and continued until about 1984.

Around 1975 three new (associate) professors refreshed the hydraulics research world in Delft: Jurjen Battjes (1939) who spoke about sea waves in his inaugural address; Jan Kalkwijk who spoke about the tides; and Matthijs de Vries (1936) on 'water and wind'. An expert in groundwater flow (which is usually attributed to the field called hydrology) was professor Arnold Verruijt (1940) who wrote a well-known textbook on the topic (1970, 1982). Battjes would become well-known for his research on the incorporation of random-wave effects in wave models for application in offshore and coastal engineering and in coastal hydrodynamics, important issues for a country which has been concerned about its sea coast for ages. Dispersion and mixing in layered flows and turbulent flows also had (and still have) the attention of the hydraulic engineers in Delft.

Professor Kramers of the Laboratory of Physical Technology in Delft (see § 3.3) would be supervisor of many PhD students. One of them was Wiero Beek (1932–2016) who finished his PhD thesis on 'mass transfer through non-permanent boundaries' in 1962. Only three years later Beek would be promoter himself, together with Kramers: in 1963 he had returned to the Laboratory as successor to Kramers,

after a short period at the Algemene Kunstzijde Unie (AKU) in Arnhem (which eventually became part of AKZO). But in 1970 Beek left Delft again, disappointed: his ideas about education were rejected and his hope for an integrated study for students in Physical Technology were frustrated.

After Beek's leaving the group was split: one for heat transfer (Hoogendoorn); and one for physical technology. The successor of Beek was a chemist, John Marriott Smith (1931–2015), who came from Manchester and would return to England in 1986.

In the early 1980s the Laboratory was confronted with severe criticism about the way it functioned: professor Rietema from TH Eindhoven (see § 4.1.2) had written an article in which he concluded that the Physical Technology groups in The Netherlands had shown no serious attempts to deepen their research activities, they just did more of the same things that they had always done. There was no progress. People like Beek agreed with this opinion but there was no general conformity about the solutions. Some thought that the use of the computer for research would mean an important step forward but others remarked that the computers and the software were still not adequate. Things changed importantly in 1988 when Harrie van den Akker (1950) became the new professor. Van den Akker, a student of Rietema, had worked for Shell as research engineer. Thanks to his efforts Shell again donated a large amount of money to the laboratory, this time not for a building or a pilot plant but for the acquisition of computers and CFD software.

In Timmerhuis (1999) Van den Akker told about the way he transformed the approach of the Kramers Laboratory and the reactions he got (translation by Fons Alkemade): "We have, to some extent, been opposed by professors in adjoining areas of science, in other faculties, and other universities. A few found that I went too much into their areas. They were not pleased. My pursuit of scientification, as I put it, the use of new measurement techniques to be able to study the processes in more detail, meant that I have been looking around in all sorts of neighboring areas. Not only in my own field of transport phenomena, but also in the broader field of chemistry, high performance computing and fluid mechanics, including turbulence and multi-phase flows. I broke out of my box and started to combine disciplines in a way which was unknown before. People who were working in fluid mechanics or high-performance computing from home, the traditional specialists in these areas, found that this was not allowed at all and was unacceptable. I do not advertise myself as an expert in fluid mechanics, but today I do a lot of work in this field. I am certainly inferior as for knowledge compared to a professor of fluid mechanics. In this sense I do really compete with the actual specialists. Thereupon I have been blown back, I had to go back into my loft."

At the end of the 1990s multi-phase flow and chemical

↑ On 26th June, 1968 the building Warmte en Stroming (Heat and Flow) of TU Eindhoven was officially inaugurated by the minister of Education and Sciences, Veringa. Here we see the main hall, 24 m high. Part of the experiments at that time were related to applications in nuclear reactors. The building also had a water tower which could provide water flows for research on two-phase flows and cavitation. In a separate building (because of the noise) high-speed gas flows could be studied. (courtesy of TU Eindhoven archives)

vapour deposition (CVD) became more important topics in the Kramers Laboratory (however, bubbly flows had been a research topic from the 1950s). Where in 1970 there had been a split, in 2002 the two groups were united again and became the section for Multi Scale Physics. In 2010 they left their characteristic building near the Department of Chemical Engineering (part of which had become known as The White Elephant) and moved to a brand-new building on a newly developed part of the TUD campus.

The Hoogendoorn group has been involved in the study of flows in spaces with thermal stratifications (e.g., during a fire in a traffic tunnel), heat transfer in both turbulent and laminar flows in complex situations (e.g., ovens), and combustion. The group also studied burners (see Ch 7).

From 1956 students could study Applied Mathematics in Delft, which was for a great deal the initiative of Rein Timman (see § 6.1). Timman and his staff would become known for their mathematical modelling, especially in the field of waves and ship hydrodynamics. From the 1970s this group also became active in numerical modelling and CFD, e.g., of collapsible tubes.

The Department of Mining Engineering was also involved in research on flow phenomena, at least in the 1980s. The topics there were separation methods for multi-phase flows (in the context of the oil industry), flows in underground formations and pits, gas flotation (separation using gas bubbles), and subterranean coal gasification.

## •4.1.2  EINDHOVEN

In 1958 TH Eindhoven, hardly two years old, had appointed Johan Slotboom (1918–1961) as professor of fluid mechanics at the Department of General Sciences which included Physics. Slotboom was an engineer from Delft who came to the NLL in the 1940s where he was head of the aerodynamical division from 1951 until 1953. Later he would have the same position at the Staatsmijnen (State Mines) in Limburg. His brother Hans became known as a key figure at Shell research.

Slotboom died when he was only 41 years old and would never see the laboratory for heat transfer and fluid mechanics for which he had made plans. In 1963 Gerrit Vossers (1926–2014) took his place, this time as professor of the new sub-department of Mathematic and Technical Physics which had started in 1960. The department of General Sciences was then led by professor Cor Zwikker, who had been director at the NLL and involved in aerodynamics. In 1962 Broer left Delft and became professor of applied physics in Eindhoven. One of his topics would be wave theory. Another theoretical physicist who became professor in Eindhoven was Daan de Vries (1916–2010). During the 1940s and 1950s he worked at the Laboratorium voor Natuur- en Weerkunde (Laboratory for Physics and Meteorology) at the Landbouwhogeschool (Agricultural University) in Wageningen. His thesis had been on heat transfer in soils, and heat transfer would also be his topic in Eindhoven from 1958. Besides heat transfer in porous media he started to study boiling phenomena, turbulent transfer of heat and mass, and transport in mixtures of gases. Whereas his colleague in Delft, Hans Kramers, was mainly concerned with 'fysische transportverschijnselen' (physical transport phenomena) related to industrial applications, De Vries would (also) focus on applications in agriculture and the environment.

In the summer of 1968 the building which was generally indicated by 'Warmte en Stroming' (Heat and Flow) was opened. Part of the building was occupied by two groups from Mechanical Engineering: Warmtetechniek (Heat Technology) and Stromingstechniek (Flow Technology). A very long walkway (typical for the TU Eindhoven campus) connected the new laboratory to the building of physical chemistry. Around 1973 the two groups of Technical Physics, Warmtetransport and Stromingsleer, became one 'vakgroep' (department) called Transportfysica (Transport Physics).

In the early years three research topics were chosen: aero-acoustics (sound production of free jets and inside tubes); the response of air bubbles to pressure changes; and thermal processes in high-speed air flows. After the 'fusion' in the 1970s three groups were formed: vortex dynamics; heterogenous media; and gas dynamics/aero-acoustics. The first topic became the specialty of GertJan van Heijst who was appointed as the successor of Vossers in 1990; Van Heijst had already worked on rotating flows with Van Wijngaarden in Twente (see below).

Physical technology got its place in Eindhoven thanks to Kees Rietema (1921–1993) who left Shell in 1959 to become professor in what he himself preferred to call 'process science'. His laboratory was opened in 1960 and he soon became well-known for his knowledge of and research on fluidized beds. When Daan Schram became lector in Eindhoven in the early 1970s (in 1979 he became professor), plasma physics became an important area there. The research included plasma turbulence. In the 1970s and 1980s the research was for a large part linked to nuclear fission but later other applications of plasmas (e.g., deposition) became prominent.

Wind tunnels have been used by some other research groups in Eindhoven: by members of the faculty of Architecture for studying wind disturbance around buildings; and by members of the Automotive group for studying the streamlining of cars. Blood flow is studied by researchers at the Cardiovascular Biomechanics group.

## • 4.1.3 TWENTE

In 1962 Leen van Wijngaarden (1932) had finished his PhD thesis on magnetohydrodynamics with Broer in Delft as supervisor. In 1965 the only two-year-old TH Twente appointed Van Wijngaarden, who was working at the NSP (MARIN) at that time, as extraordinary professor of 'warmte- en stromingsleer' (heat transfer and fluid mechanics) at the Department of Mechanical Engineering (there was no Physics Department yet). In his inaugural lecture of January 1966 Van Wijngaarden noted for his audience that in the Netherlands fluid mechanics was only taught at the technical universities. He mentioned an article by Wayland Griffith, scientific director of Lockheed, which had appeared two years earlier in the Journal of Fluid Mechanics (founded in 1956) entitled 'Is fluid mechanics becoming extinct as a branch of science?'. Griffith had written about one of the main differences between fluid mechanics and other branches

of physics: "... as in other areas certain discoveries lead to technical applications, fluid mechanics is usually aimed at getting insight into and deepen knowledge about situations which have already been observed in technology. ... [Only seldom] a brand new technological concept has been made possible by a basic discovery about the nature of fluid flow." Van Wijngaarden tried to convince his audience that fluid mechanics did have a future. In 1962 Pieter Zandbergen (1933-2018) also finished his PhD thesis on supersonic flow around bodies in Delft, under the guidance of Egbert van Spiegel (1927) at the Department of Aeronautical Engineering. At the same time as Van Wijngaarden, Zandbergen was appointed as professor in Twente, to lecture on applied mathematics. TH Twente was still taking shape in the late 1960s. In 1968 Zandbergen was one of the founders of the new Department of Applied Mathematics and in 1969 Van Wijngaarden would also be professor in the newly established Faculty of Applied Physics. This faculty had been forced, like all physics departments, to make a choice between the many fields in physics and had decided to go for phenomenological physics, which included fluid mechanics, but to exclude high energy physics and reactor engineering. It was decided then that the Fluid Dynamics group would become part of both the Physics and the Mechanical Engineering Department.

From the time Van Wijngaarden started with his group in Twente they worked on gas bubbles in liquids, e.g., on pressure waves in bubbly flows. The approach was initially theoretical, but then later also extended to experimental and numerical. This can be explained from the cavitation phenomena which he had observed and studied during his time in Wageningen. Other topics were: surface waves, standing waves in tubes, rotating flows, and cavitation. When Van Wijngaarden retired he was succeeded by two professors: Harry Hoeijmakers (Mechanical Engineering) and Detlef Lohse (Applied Physics).

From 1971 TH Twente also had a chair for rheology, occupied by Piet van der Wallen Mijnlieff (1924–1996), a chemist whose PhD thesis had been on colloidal electrolytes and who had worked at the KSLA. For several years he was chairman of the Nederlandse Reologische Vereniging (Dutch Society of Rheology), founded in 1951. His successor was Jorrit Mellema (1943–2009), a physicist, who was one of the authors of a well-known text-book on rheology and rheometrics: Inleiding in de reologie.

As in Delft, at the Twente University a section for tribology had been founded decades earlier. One of the topics since the 1960s has been elasto-hydrodynamical lubrication: lubrication in which significant elastic deformation of the surfaces takes place which considerably alters the shape and thickness of the separating lubricant film. One of the 'hydrodynamical' PhD theses which was produced from this section was entitled Stokes flow in thin films (2001).

Twente also has a tradition in hydraulics research. Today, the two chair groups: Water Engineering & Management; and

Marine and Fluvial Systems, are both part of the Twente Water Centre. Head of the first group is Suzanne Hulscher; in 2002 she became – probably – the very first female professor in fluid mechanics in The Netherlands. She holds the chair for Marine and Fluvial Systems.

## • 4.1.4 OTHER UNIVERSITIES

Fluid mechanics research in the Dutch academical world is for a large part research taking place at the technical universities. Other universities never had such a diversity of research topics related to fluid mechanics, but some must be mentioned here for their specific expertise.

### GRONINGEN
Groningen has a long tradition in numerical fluid mechanics which started in 1958 when Adri van de Vooren (1919–2007) became professor in Applied Mathematics. Van de Vooren had been trained as a physicist in Delft, had become a specialist in flutter at the NLR and was extraordinary professor in aero-elastics in Delft from 1956. In Groningen he was one of the people who successfully pleaded for the acquisition of one of the first commercial Dutch computers, the ZEBRA, which came to Groningen in 1959 (two years after one had been put into operation in Delft). Van de Vooren was one of the founding fathers of the new field of Technical Mathematics in Groningen and of the CFD activities there. Another founding father was Joop Sparenberg (1924–2010) who had been trained as an aeronautical engineer and had worked at the NSP. Sparenberg and Van de Vooren also managed to have the university of Groningen open an engineering field of study: Technical Mechanics. Several PhD theses on ship propulsion have been written in Groningen.

### LEIDEN
At the University of Leiden interest in the behaviour of fluids had been greatly limited to super fluids, the research topic which had become famous due to the laboratory of Kamerlingh Onnes. After the Second World War this interest continued as is evidenced by the title of the PhD thesis of Peter Mazur (1922–2001) which was published in 1951: Thermodynamics of Transport Phenomena in Liquid Helium-2. Mazur became an associate professor at Leiden University in 1954. In 1955, he and his thesis advisor Sybren de Groot (1916–1994) founded the Lorentz Institute for Theoretical Physics at Leiden University. In 1961, Mazur became a full professor, and when de Groot left in 1963, he became director of the Institute. Besides nonequilibrium thermodynamics, for which he and De Groot became famous, Mazur would also initiate research into fundamental problems of fluid mechanics. To describe diffusion of large particles in fluids, in 1974 he introduced, together with Dick Bedeaux, the concept of induced forces. This concept was used to develop a theory for the viscosity of a suspension. Around

1982 Mazur, Wim van Saarloos (who would later become director of FOM and president of the KNAW), and Carlo Beenakker developed an algebraic method to successfully describe hydrodynamic interactions between arbitrary numbers of particles using induced forces. This was a breakthrough in the field. In more recent years, physicists in Leiden have published papers on viscoelastic flows and on the behaviour of granular materials.

### UTRECHT
As we have seen (§ 2.2) the national meteorological institute KNMI was founded in Utrecht, and it will not be surprising that the University of Utrecht has been involved in meteorology for many years. From 1966 it housed the Institute for Meteorology and Oceanography, which organized lectures on fluid mechanics for students (one of the teachers was Frans Nieuwstadt, later GertJan van Heijst got this position). From 1991 the institute (which in 1987 had moved from the faculty of Geosciences to that of Physics and Astronomy) was known as IMAU: Institute for Marine and Atmospheric research Utrecht. One of the professors in Fluid Mechanics was Leo van Rijn (1946) who had worked at the WL. One of the present weathermen on Dutch national broadcasting (NOS) had been a student at IMAU. Colloids had a long tradition in Utrecht which started with Hugo Kruyt (1882–1959), with whom Burgers had had close contacts when they were both member of the Viscosity Commission of the KNAW (see § 5.3) and on issues concerning science and society. Later the colloid group in Utrecht would be led by professor Henk Lekkerkerker.

### WAGENINGEN
Willem van Wijk (1906–1967) was trained as a physicist in Utrecht, worked at BPM in the 1930s and 1940s and became professor at the Agricultural University in Wageningen in 1947, where he started to introduce the 'physical method' for the solution of technical problems related to agriculture. He was one of the initiators of the Landbouw Physisch Technische Dienst (Agricultural Physics Technical Service) around 1955. This organisation built several machines related to fluid mechanics, such as a water drainage meter and a measuring instrument for the mobility of sperm. After Van Wijk's death one of the topics of research in his group became transport phenomena (including turbulence) related to SPAC: the Soil Plant Atmosphere Continuum.
The Kraijenhoff van de Leur Laboratory for Water and Sediment Dynamics has been named after Dirk Kraijenhoff van de Leur (1918-2001) who became professor of 'hydraulics and catchment hydrology' in Wageningen in 1966 and had opened his own laboratory in 1965 (see also § 6.2.3); from 1971 there was a close collaboration with the WL in Delft. Meteorology had been a topic in Wageningen since the start of this university in 1918. From 1962 Hans Lyklema (1930–2017) made Wageningen also known for its research on colloids.

# 4.2 INSTITUTES

As we have seen in § 3.2 before the Second World War three important research institutes closely related to fluid mechanics had been founded: RSL (later NLL and today NLR); WL (later Delft Hydraulics and today Deltares); and NSP/NSMB (today MARIN).

After 1945 other institutes, some of them founded decades earlier, started to take up (fundamental) research in fluid mechanics. One of them was the KNMI which had been mainly involved in measuring and in developing measuring instruments and procedures before the war. In the 1960s the section Physical Meteorology was set up in which one of the main themes would become the atmospheric boundary layer (see Ch 7) and the generation of waves by wind. Another institute with a long history (1876) is the Netherlands Institute for Sea Research (NIOZ). In 1957, its director proposed to broaden their scope from biology to the four pillars of oceanography: biology, chemistry, physics, and geology. His proposal was instantly approved. From 1966 the NIOZ was involved in an ambitious research program for the Noordzee (North Sea), in which also RWS, the WL and the KNMI were involved, among others.

Other institutes which have a link with fluid mechanics did not exist before the war. The CWI (Center Mathematics and Computer Science) was founded as the Mathematisch Instituut in 1946 and would soon become involved in the development of the first computers in The Netherlands. In later years CWI research took up CFD, e.g., for the simulation of high-speed gas flows and of free surface flows of water (e.g., around ships). In 1955 the Energy Research Centre of The Netherlands (ECN) was founded, initially for research on nuclear energy and related topics. Around 1975 the Dutch government wanted to stimulate wind energy and in the National Research Program Wind Energy of 1976 ECN took an important role (see also Ch 7). Initially the research was mainly aimed at measuring performance, but the research (performed by the Flow Energy section) later became more fundamental. Together with the Stork company, airplane builder Fokker, TNO (see below), and others ECN started to develop new wind turbines. In the 1990s ECN developed simulation software which builders of wind turbines could use to improve their products. Despite a disappointing growth of the number of wind turbines in the Netherlands and a decrease in the number of builders, ECN managed to gain a leading world-wide position in rotor aerodynamics.

## •4.2.1 TNO

A somewhat special place in the Dutch landscape of scientific research is taken by TNO, the organisation for Toegepast Natuurwetenschappelijk Onderzoek (Applied Scientific Research). Already during the First World War (when The

Netherlands was confronted with shortages of fuels, raw materials, etc.), the idea arose to establish a national institute where social problems could be investigated by means of scientific research. It took the implementation of a special law in 1930 before TNO was officially founded in 1932, but due to a continuing debate on what TNO should be and do, and due to the War, this institute would only start to flourish from 1945. Several TNO laboratories were set up, initially most of them in and around Delft due to a strong connection with the TH. The NSP in Wageningen, the WL in Delft, and the NLL in Amsterdam all became TNO institutes after the War, but this lasted only a short period. Former NSP director Troost became one of the TNO directors after his return from the USA in 1960. ECN became part of TNO in 2018. In several TNO laboratories research related to fluid mechanics has been, and still is, carried out. In one of the laboratories for military research, for example, ballistic studies have led to flow models and simulations. At the TNO institute TPD in Delft aero-acoustic research has been done. One of the TNO institutes where fluid mechanics became a major research topic was that in Apeldoorn. Some years after its official opening in 1974, a water tunnel and a boundary layer wind tunnel were realized; the first has never been successful and was later transformed into a wind tunnel. The boundary layer tunnel has been successfully used for forty years to study wind disturbance around buildings and other objects. TNO Apeldoorn was closed in 2014. Today most TNO expertise in fluid dynamics (and heat transfer) is situated in Delft. Fields of expertise of TNO include pipeline systems, building physics, flow chemistry and flow control.

## •4.2.2 WL / DELFT HYDRAULICS / DELTARES

After the Second World War the WL had plenty of orders, also from abroad (known internationally as the Delft Hydraulics Laboratory). Thijsse got concerned about the lack of real research ('speurwerk') in his laboratory. He also saw that in the USA the computer had started to become an interesting tool for engineers (see also § 6.3). In 1960 the WL got more than 700 requests for model research, most of them from RWS and many of them related to the Delta Works and the plans for Europoort (Rotterdam Harbour extension). From 1970 the growing interest in environmental issues became clear: investigations on the spreading of pollutants in seas and rivers were undertaken. Among the 'traditional' topics of research were riverbeds, coastal erosion, dredging, estuaries, river morphology, sediments, tidal currents, and waves. At the WL experiments on ship models were also performed, usually related to the behaviour of ships in harbours (mooring) and waterways.

In 1960 Thijsse was succeeded by Schoemaker. The number of fundamental research projects grew, which led to several

↑On the outskirts of Delft, close to the TH campus which had been erected from the 1950s in the Wippolder, the new WL was built from 1965 onwards. The area was called Thijsse's erf (Thijsse's courtyward) by the WL people; the official address today is Boussinesq-weg 1 (named after a famous French hydrodynamicist). In the middle, with the tower, is the main building with the staff offices. On the upper left we see the Stromingslab (Flow lab) which has now been demolished. On the lower left is the Windgotengebouw (Wind flume building); other flumes are in the building to the right of the main building. The huge hall to the right of the Stromingslab was called the Zout-zoethal (Salt-fresh hall). (courtesy of Deltares)

PhD theses related to the WL. In 1965 the first part of the new WL, south of Delft, was realized. In 1969 the new site of the WL was officially opened, including the new and unique wave flume. During the 1970s several offices and halls were added, even after the Queen had opened the new WL officially in 1973.

The old building near the city centre became the place for research on 'pumping and industrial circulation'; one of the topics there were dredging pumps. At the location De Voorst, the 'open air laboratory' concessions were made: several huge halls appeared in which the hydraulic engineers could do their job without being disturbed by bad weather.

From the 1980s the institute became known as Delft Hydraulics. The WL has been working with engineers from the Rijkswaterstaat since its foundation (see § 3.2). In the 1980s RWS had nine 'directies' of which Waterhuishouding and Waterbeweging (DWW) (water maintenance and movement) was one. Part of the DWW and other departments of the RWS merged with Delft Hydraulics in 2008 to become Deltares.

• 4.2.3 NSP / NSMB / MARIN

From the 1950s several developments took place in shipbuilding and shipping: tankers became much larger and heavier, speeds and engine performances of ships increased, vibrations and noise became more important. This led to a significant growth in research and test orders for the NSP in Wageningen; and to a more scientific approach. Topics were (and still are) cavitation, propulsion and resistance of ships, and ship motions in waves. Thousands of ship models have been built at the institute. The up-and-coming offshore activities worldwide, from the early 1960s, also led to an increase in the research activities. A Wave and Current Basin was built in 1973 (which was replaced by a complete new Offshore Basin in 2000). Another important development in the 1960s was the use of computers and the opening of a Computer Centre in 1961. Increasingly, the institute would also become a centre of knowledge on numerical tools for ship and propeller design.

Van Lammeren, director from 1952 till 1972, managed to let

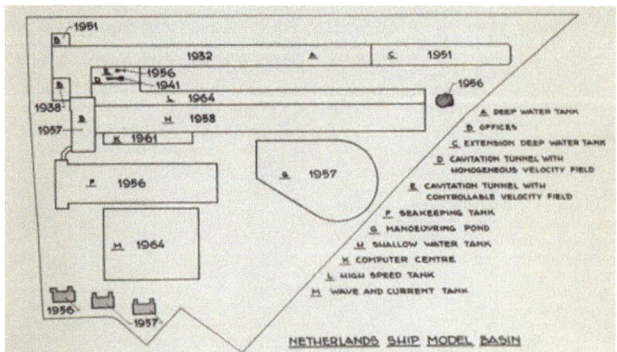

↑From about 1955 an impressive number of new tanks (also called basins) and other facilities was added to the NSP. The reason that the NSP was founded in Wageningen – not really a centre of activity in ship building – had to do with two circumstances: the fiancée of its first director, Troost, lived there and the soil structure in Wageningen was much more fit to support a heavy towing tank than that in the western part of The Netherlands. (courtesy of MARIN / from: Van Lammeren (1963))

his institute prosper. Shortly before he retired, an important new facility was added in the nearby city of Ede: the depressurized towing tank (see § 6.2). Shortly after his retirement, a crisis in the international shipping world arose but the NSP managed to survive. Reduction of ship resistance became even more important and led to new developments in ship design (see also § 7.2). In 1980 its name was changed to MARIN (from Maritime Research Institute Netherlands). Today, MARIN has close ties with the Wageningen University.

### • 4.2.4 NLL / NLR

As we have seen in § 3.2.1 times had been tough for the NLL in the decade after the Second World War. In 1949 there were still 304 employees, one year later there were 204. For many decades, this shock effect led to a very cautious hiring policy for permanent personnel. Nevertheless, by 1983 the NLR had 700 employees and the number was still growing. From about 1960 income also started to grow significantly, while the government subsidy remained almost constant at about 30% of the total income.
In 1946 a new section was set up at the NLL, the department for flutter and theoretical aerodynamics (there had been an Aerodynamics Department for many years and this would be continued alongside the new department). Flutter is the dynamic instability of an elastic structure in a fluid flow and had already been the object of study before the War. One of the experts in this field at the NLL was Johan Greidanus (1911–1993) who had been trained as a theoretical physicist in Amsterdam. Greidanus can be regarded as

an important mentor for people like Timman and Van de Vooren, mentioned above. Zandbergen was also educated in this section. In 1951 Greidanus became head of the aerodynamics department of Fokker, the Dutch airplane builder. In 1958 the NLR had its first digital computer, the Dutch ZEBRA which was also bought by the University of Groningen. With its advent, the Computation Office of the NLR was changed to the department of Mathematical Problems and Numerical Calculations, headed by Van Spiegel (see also § 4.1.3). At about the same time two other new departments were added, showing the growing interest in high-speed flight and rocket science: Transonic Aerodynamics and Gas Dynamics. The involvement with space flight would also become apparent from the change of name of the institute in 1961: from NLL to NLR (where the R is for 'ruimtevaart' (space travel)). Today the original meaning of the abbreviation has become irrelevant and the institute, which has become more commercial in the last few decades, today is presented as the Netherlands Aeropsace Centre (NAC).

## 4.3 INDUSTRY

As we have seen in § 3.3, some Dutch companies already had research departments before the Second World War but for most of them serious research only started after 1945. Especially during the 1950s Dutch industry became 'research-minded', whereby the more fundamental kind of research (or 'free research' as it was then called) not directly linked to a product or a production process, also got serious attention. Especially in the 1960s, with a booming economy,

←↑Since 1940 the main building with the main entrance (on the right) of the NLR in Amsterdam has hardly changed from the outside. It was the designed by the architect Hugh Maaskant and was completed shortly after the German invasion in May. Today it is a national monument. Integrated with the building were two wind tunnel halls where research could be done during the war years (till December 1944 when the supply of electricity was largely reduced). The north side of the NLR premises today is much less attractive. On the other hand, this side can be reached via the Windtunnelkade, one of the very few streets in the world including 'wind tunnel' in its name.

↑By 1963 the Physical Laboratory of Stork had become an important part of this construction company (engines, turbines, pumps, boilers, etc.). This photo is from a presentation album showing many of the activities at the lab. Here profiles of fans are investigated in a subsonic wind tunnel. (courtesy of Historisch Centrum Overijssel / Fotocollectie Stork)

the budget for research seemed to be unlimited. Some companies already had a longer tradition (e.g., Philips, Shell, Werkspoor) but others (e.g., Unilever, Hoogovens, AKZO) had little experience. In some branches of the industry research was considered normal, in others it was not very common, at least before the Second World War (as in the case of construction companies like Stork and Werkspoor). Shell played an important role in fluid mechanics in The Netherlands in general. In Amsterdam and Rijswijk numerous investigations were done, particularly on multi-phase flows, but not all the results were published. More important was the financial support which Shell provided to the academic world. Van Wijngaarden in Twente, for example, received money for his research from Shell for more than ten years. Several of his Masters students found employment with Shell.

In the 1990s dark clouds showed up for Dutch industrial research. Professor Hoogendoorn (see § 4.1.1) remarked about the reduction of research budgets in his valedictory lecture of 1998: "For the future I see an important growth of CFD [computer simulations]. A point of concern for me is the fact that in some sectors of industry due to shrinkage in the R & D organisation the in-house expertise with regard to physical fluid mechanics tends to disappear. This is dangerous, before you know it you make the wrong calculation."

Below one finds a short and non-exhaustive description of three branches of Dutch industry. In the chapters 7 up to 13 the reader will find several cases of industrial research.

## UNILEVER

Research at Unilever had already started during the Second World War in its laboratory in Zwijndrecht, but the opening of the Unilever Research Laboratory in Vlaardingen in 1956 by prime minister Willem Drees meant a big step forward. Around 1956 there were about a hundred researchers, ten years later there were about a thousand. In the beginning the main topics in Vlaardingen were edible oils and fats; later detergents and chemical products also became important. Though a large part of the Unilever research was, and is, related to chemistry, physical processes were studied from the 1960s by one of the research groups which were doing purely scientific work, the Physical Chemistry group. One of the topics of this group was the rheology of multi-phase systems. In 1962 a PhD thesis was published by one of the Unilever (female) researchers, entitled: Stabilization of water in oil emulsions by solid particles. In De Ingenieur of 1965 another researcher wrote on "the flow image of a moderately viscous fluid in a vortex chamber atomizer". More on the food research at Unilever can be found in Ch 7.

## DSM

As we have seen in § 3.3 the State Mines in Limburg were taking scientific research quite seriously from the 1920s and this resulted in the foundation of a Centraal Proefstation (Central Pilot Plant) in 1929 where several 'practical' aspects of the production process (ventilation, washing of coals, etc.) were investigated. In 1940 the research activities were broadened with the opening of the Centraal Laboratorium (Central Laboratory). An important part of the research was now related to chemical processes since coal was used to make several chemical products. After the Second World War coal remained an important fuel for The Netherlands until huge gas fields were discovered in the North of the

↑ In the severely protected cellar of the VSL building in Delft (see also § 6.2.1) one finds these intriguing cylinders. They are so-called Active Piston Provers, used for the calibration of flow meters. The APP was developed by the Bronkhorst company from Ruurlo. Bronkhorst has been active in developing flow meters since the early 1980s and became famous for its meters which make use of the Coriolis effect.

country. In 1973, when the last mine was closed, the name of the company was changed to DSM. By then DSM had already changed into a producer of chemical products. One of these were (strong) polymer fibres for which the gel spinning technique was developed.

### FLOW METERS AND CALIBRATION
A rather special branch of industry related to fluid mechanics is that for the design, production, and calibration of instruments for measuring fluid flows. For many years most researchers and engineers had to deal with instruments involving propellers which had hardly seen any innovation from the pre-war years. It seems that in The Netherlands serious flow meter development only started in the 1960s. In 1966 the Dutch branch of the German Krohne company finished the world's largest calibration rig for volume flow in Sliedrecht, the home town of the Dutch dredging industry. In 1974, engineering student Anton van Putten pioneered the

development of MEMS flow sensor technology by implementing the world's first thermal mass flow sensor on a silicon substrate. This flow sensor would become the mother of all CMOS flow sensors and its revolutionary design can still be found in many other flow sensors around the world. In 1998 he started his own company, VP Instruments.

## 4.4 BRINGING TOGETHER THE THREE WORLDS

After the Second World War the need for engineers became huge since the country had to be reconstructed. Industrialisation and mechanisation got an enormous boost and the TH Delft was supposed to deliver the young engineers who had to do the job. As for fluid mechanics, hydraulics was given a lot of attention, especially after the Storm Surge of 1953. But apart from 'applied fluid mechanics' the more fundamental side of the field was certainly not neglected. The 'Colloquium over turbulentie' which was held in Delft in 1950 attracted about fifty participants from universities, institutes, and industry; some years later a second edition took place.

For those working in fluid mechanics in The Netherlands, there were only very few national meetings to discuss their work. Every four years the International Congresses on Theoretical and Applied Mechanics were held and that was (more or less) the only possibility for Dutch fluid mechanicists to meet their fellow researchers, including those from other European countries and from outside Europe. The only national meetings on mechanics for engineers (and others) were organized by the Royal National Society of Engineers (KIVI). KIVI also organized 'leergangen', multi-day courses during which a specific topic was treated by specialists (also from universities) for an audience of engineers working in the field.

For physicists working in fluid mechanics, the situation can be considered as a lot worse. During the Second World War scarcely anything happened in Dutch physics research and it consequently lost its leading international position. Things quickly changed after the liberation of the Netherlands. The need for new research institutes, which had already been felt among scientists and politicians since the 1930s, had strongly increased. That was partly due to the highly promising developments for nuclear energy and the increased awareness that other countries had acquired a stronger position in the physics arena.

Leading physicists and the government joined forces in 1946 to give fundamental physics a boost. In April of that year they established the Foundation for Fundamental Research on Matter (FOM). The founders chose the name 'Research on Matter' to convey their broad vision: besides (nuclear) physics they also wanted to build a bridge to materials science, chemistry and other disciplines. In its initial years, FOM focused mainly on nuclear physics, mass separation and analysis, and metallurgy. In the first decade after it was founded, FOM expanded its research activities to include solid-state physics, plasma physics, and high-energy physics.

In the 1960s FOM experienced a stagnation in the growth of the organisation. Cutbacks forced the foundation to set priorities and to adopt a strict policy for the assessment of research proposals. At the same time, academic groups found it almost impossible to get financial support from the industry. From about 1970 FOM adopted a new spearhead in its policy: industrial innovation. At the end of the 1970s the programme Applied Physics and Innovation was set up to accommodate this. In 1981 this programme served as a model for the establishment of Technology Foundation STW. Around 1980 fluid mechanics was still not a field of physics which had the attention of FOM. In a two volume publication: Physics in The Netherlands – a selection of Dutch contribu-

tions to physics in the first 30 years after the second world war, (published by FOM in 1982); a lot of physical topics were treated but not the physics of fluids. Only some related areas were described: the hydrodynamics of superfluid helium (some scientists at Philips Research Laboratories published about vortex rings in liquid helium II in the 1960s); and magnetohydrodynamics. In the *Nederlands Tijdschrift voor Natuurkunde* (Dutch Journal for Physics) of the Nederlandse Natuurkundige Vereniging (Netherlands' Physical Society, NNV), of which Burgers had been one of the founders, fluid mechanics still didn't get much attention during the 1980s. In 1984 a report on the status and the perspectives of physics in The Netherlands (Heijn et al., (1984)) revealed that Dutch physics scored above average internationally. It also suggested to FOM the installation of workgroups dedicated to fluid mechanics, a field of physics that had had very little attention from the FOM policy makers. It was probably due to this policy, the committee wrote, that fluid mechanics in The Netherlands stayed behind compared to the rest of the world. From the report: "Some research groups have found financial support from the 'derde geldstroom' [finance from industry and institutes; FA]; this stimulates the orientation towards applications but not always the internal development in quality."

The early 1980s were harsh times for science at the universities: budgets were diminished and professors had to find finances outside their university. To illustrate the consequences of the deteriorating circumstances: Van Wijngaarden had to decide to have his supersonic wind tunnel and his flume for wave research removed from his laboratory.

Dissatisfied with FOM's policy, three professors of fluid mechanics (Van Wijngaarden, Vossers, and Hoogendoorn) started to address the issue, and with some success as in 1987 the 'Stimulation Programme for Fluid Dynamics and Heat Transfer' was started by FOM. In the next year the various research groups of the TU Delft working in fluid mechanics had realized the Samenwerkingsverband Stromingsleer Delft (Partnership Fluid Mechanics Delft); its promoters were Van Ingen, Nieuwstadt, and Hoogendoorn. This partnership can be considered as a forerunner of the Burgers Centre. The three technical universities which now had started to coordinate the fluid mechanics research in The Netherlands decided to give each of them a key topic: turbulence for Delft; rotating flows for Eindhoven; and multi-phase flows for Twente. Rheology would be present in the program of each of them.

Vossers recalled in his farewell lecture of 1989: "For the field of flow and heat I have tried, for the last few years, to realize a regular contact between the different 'vakgroepen' (working groups) in The Netherlands. This does not automatically happen via the 'second cash flow' organizations [the money from the public organizations for the promotion of science] in which sometimes 'werkgemeenschappen' (working communities) fulfil this function. This is especially the case when there is inadequate attention for a field of science at the national level. Curiously, in contrast to the USA, England, and France, the Dutch physics community didn't regard fluid mechanics of sufficient interest; in this situation some change can be detected now after long talks." Two years after Vossers's speech the 'Werkgemeenschap voor Strom-ing en Warmte' was established (also thanks to the effort of Nieuwstadt) and got its own budget. At the end of the 1990s the Werkgemeenschap was dissolved due to a reorganisa-tion of the FOM research policy.

A very important event for fluid mechanics in The Nether-lands was the foundation of the national research school ('onderzoeksschool') named after J.M. Burgers, in 1992. In the early 1990s the Dutch government had stimulated the cooperation of groups in the same field of science and the Burgers Centre was among the first to get recognition.

This meant that money would come from NWO (the Dutch Science Organisation, of which FOM was a suborganisation for physics) and from the Ministry of Economic Affairs. Fluid mechanics benefited from the growing interest in 'applied science' with the Dutch government: whereas in fields like plasma physics it was hard to point out the applications, for fluid mechanics this was much easier. Besides, flow phe-nomena could usually be visualized.

Led by its scientific director, Charles Hoogendoorn, the main goals of the J.M. Burgers Centre were formulated thus:
- the coordination of research and educational programs on fluid mechanics;
- the stimulation of cooperation among the participating groups;
- the organisation of an educational program for PhD stu-dents;
- the stimulation of cooperation between the academic world on the one hand and the groups from industry and institutes on the other hand;

## FLUID MECHANICISTS AS SENIOR OFFICERS IN UNIVERSITIES, INDUSTRY AND GOVERNMENT

As one browses through the history of fluid mechanics in The Netherlands, one encounters quite a lot of men (fluid mechanics has seen few female engineers in the past hundred years) who have not only earned their stripes for their work as fluid mechanics experts but also as senior officers. This may lead to the question whether people trained in fluid mechanics are especially well fitted to exercise managerial roles, but we will not dwell on this question here. We just give some examples.

Gerrit Vossers (see § 4.1.2) was a student of Naval Archi-tecture in Delft from 1945 till 1955. He published his PhD thesis on 'slender body theory in ship hydrodynamics' in 1962 under the guidance of Timman. He worked at the NSP (MARIN) for some time but in 1962, at 35 years old, he was appointed professor of fluid mechanics at the TH in Eindhoven, at the sub-department of applied physics. In 1970 he became chairman of that sub-department and only one year later he became rector magnificus of the TU. The early 1970s were quite 'turbulent' at the Dutch universities due to students' protests and administrative reforms. Vossers certainly did not have an easy job but he managed well. After he left the rectorate in 1976, he remained active as a mediator and advisor for reorgani-sations in the academic world.

In those same early 1970s the rector magnificus of the

TH Twente was Pieter (or just Piet) Zandbergen (see also § 4.1.3), who had studied aeronautics in Delft and started his career at the NLR. In his PhD he treated supersonic flow around bodies. In 1966 he became professor in Twente where he wrote several important papers, e.g., on Von Kármán swirling flows and on numerical simulation of nonlinear water waves (in co-operation with research-ers from MARIN and Delft Hydraulics). The managerial skills of Zandbergen had been noticed and in 1996 he was elected president of the Royal Netherlands Academy of Arts and Sciences (KNAW) for a period of three years. Zandbergen left the rectorate in Twente in 1974 and eight years later another professor involved in fluid mechanics would take on this job. Wiebe Draijer (1924–2007) had been a student of mechanical engineering in Delft and would become professor in 'industrial heat technology' at the TH Twente in 1964. Draijer was a member of the Eerste Kamer (Senate) from 1971 till 1974 and became involved in development cooperation.

As has been mentioned in § 4.1.3 Leen van Wijngaarden became professor in Twente at about the same time as Zandbergen. He had been a student of Burgers in the last few years while Burgers was still in Delft and was a research associate at AHD from 1959. Like Vossers, Van Wijngaarden started his career at the NSP. In 1984 he en-

- the strengthening of contacts between the Dutch academic groups and similar groups in the rest of the world.

In the first year thirteen groups became part of the Burgers Centre, involving about thirty professors. Five main research themes had been chosen:
- turbulence and non-linear dynamics of flows;
- multi-phase flow, disperse systems and rheology;
- free surface flows and waves;
- computational fluid dynamics and modelling;
- compressible fluid dynamics.

One could say that the Burgers Centre definitely put fluid mechanics on the map of Dutch research. But in 1997 it became evident that still not all physicists regarded fluid mechanics as belonging to 'real physics'. In January 1997 physicist Ad Lagendijk wrote in his weekly (polemic) column in a national newspaper that in his view technical universities were not real universities, their research was too much 'applied'. This provocation caused several letters to the editor of the same newspaper. One of those letters was written by Henk Ten-

nekes, then professor of meteorology at the VU in Amsterdam and aeronautical engineer from Delft.

"Does little Ad really think that one can design a safe plane using that kind of nonsense? Does he really think that engineering work is just a licky extension of physics, job creation for third-rate citizens who cannot keep up with the bluff poker of the super-profs? Little Ad Lagendijk is the personification of the idiotic privileges which the national physicists claim for themselves." [translation by FA] Lagendijk reacted one week later, remarking that his words had been misinterpreted: "I do not consider technology as inferior and I have also never said this. The only thing I find, is that technology and science are different affairs."

Today the Burgers Centre counts 25 academic research groups, involving almost fifty professors. The groups have close and fruitful contacts with the important industrial groups working on fluid mechanical topics and with the research institutes. It is alive and kicking and critical comments like those of Ad Lagendijk haven't been heard for many years.

tered the Bureau of IUTAM as treasurer and later became president (1992–1996) and vice-president of that important international organisation where he would become acquainted with many fluid mechanics experts from all parts of the world. Van Wijngaarden also got involved in the activities of EUROMECH, the European Mechanics Society, of which he became a honorary member. He has also been an editor of the famous Journal of Fluid Mechanics (JFM) and of Fluid Dynamics Research. As we have seen in §4.4 Van Wijngaarden was also one of those who woke FOM up to fluid mechanics and was one of the founding fathers of the Burgers Centre.

Usually, professors in fluid mechanics come from institutes or industry (especially Shell) but it is seldom that they leave the academic world to start a career in industry. One of those who did so, was Wiero Beek (see §4.1.1). In 1970 he became director of Unilever Research (see § 4.3). During his 25 year stay there Beek also found time for several other activities, one of them being the foundation of a 'network' for discussing the human threats to the natural environment and the possible (technical) solutions for these. This work led to his membership of the Wetenschappelijke Raad voor het Regeringsbeleid (Scientific Council for Government Policy), an independent think-tank for the Dutch government.

↑Egbert van Spiegel wrote his PhD thesis under the guidance of Timman, worked at NLR and became professor in Delft himself in 1960, at 33 years old. He was one of the first who did research, with a PhD student, on numerical problems related to flow equations, in 1964. Later he would leave the academic world and became director-general of science policy for the Dutch government. (courtesy of TU Delft / photo by Fotografische Dienst via Beeldbank / CC BY)

© Springer Nature Switzerland AG 2019
F. Alkemade, *A Century of Fluid Mechanics in The Netherlands*, https://doi.org/10.1007/978-3-030-03586-0_5

# RESEARCH IN FLUID MECHANICS: FLOWS

05

Fluid mechanics, like many other fields of mechanics, has two faces: the fundamental one and the applied one. Fundamental research is mainly done at universities, much of the applied research can be found in institutes and in industry. But this division is not strict and especially in the 1950s, 1960s and 1970s fundamental research could also be found outside the academic world. As for the more fundamental studies, all kinds of flows been in the focal point and in this chapter for several of these examples from Dutch research are presented. The flows have been subdivided among three categories: single-phase flows, flows related to the field of rheology, and two-phase flows. Today, the fluids and flows in the last two categories are also indicated as 'complex fluids' and 'complex flows'.

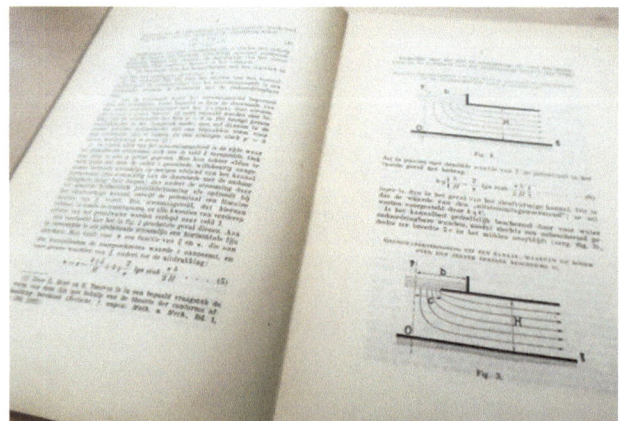

↑In the 1910s and 1920s much was written about groundwater flows in the Netherlands (see also § 6.1). The only paper by Burgers on this topic was published in 1926 in *De Ingenieur*, the periodical of the Royal Netherlands Society of Engineers (KIVI) which at that time was still more like a scientific journal. Note the drawings, indicating the assumption of a 'smooth' (laminar) flow. (courtesy of Burgers Archives / TU Delft)

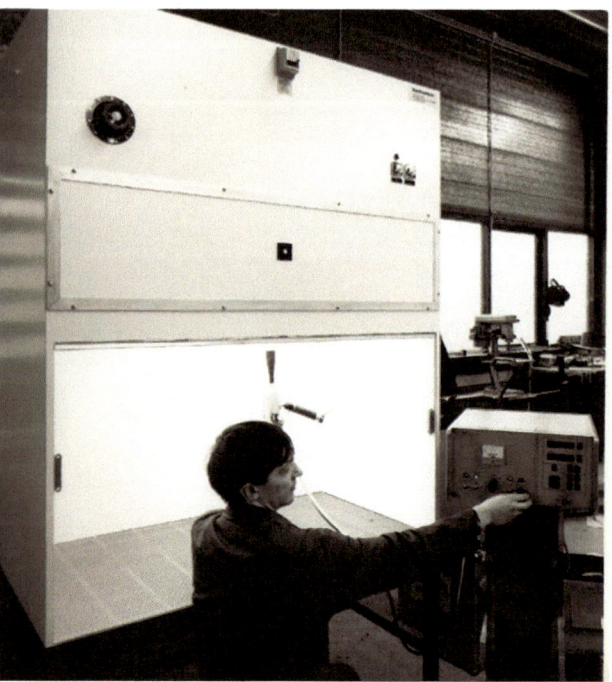

↑Making laminar air flows has been one of the specialities of the Dutch Interflow company for more than forty years. Among their products are laminar flow cabinets, carefully enclosed benches designed to prevent contamination of semiconductor wafers, biological samples, or any particle sensitive materials. Air is drawn through a filter and blown in a very smooth, laminar manner towards the user. Due to the direction of air flow, "the sample is protected from the user, but the user is not protected from the sample". (courtesy of Interflow)

# 5.1 SINGLE-PHASE FLOWS

The flows in this category have been subdivided among seven subcategories, for reasons of convenience and clarity. For each category examples of research from the 'early period' (many involving work by Burgers and co-workers), and also of more recent times are presented.

## • 5.1.1 LAMINAR FLOW

Laminar flows were for centuries the only flows that could be and were studied by scientists working on hydrodynamics. During the 18th and 19th centuries the theory of potential flows was developed: flows for which the velocity field is the gradient of a scalar function. These fields show no rotational behaviour (no vorticity) and could be called 'smooth'.
The study of turbulence started rather late. The transition from laminar to turbulent behaviour was investigated experimentally by Reynolds in Manchester in the 1880s, who still used the term 'sinuous' instead of 'turbulent'. This generated some interest in turbulence, but little was done particularly on the theoretical side. The reason for the almost complete

absence of a theoretical approach to turbulence was simple: up until about 1920 hardly anyone had an idea of how to handle the Navier-Stokes equations regarding turbulent flows. It was only under certain assumptions that these equations became manageable. One example of this is the assumption that the inertial forces in the flow are small compared to the viscous forces (i.e., the Reynolds number Re is very small). In this case the Navier-Stokes equations can be linearized and become the Stokes equations; its solutions are called Stokes flows (or creeping flows).
As we have seen in § 2.4 Lorentz found a solution for the Stokes equations in 1896. In his paper Lorentz came up with what is now called the Lorentz reciprocal theorem. This is a very general theorem in low Reynolds number hydrodynamics, relating the stress and velocity fields on an arbitrary surface. The theorem is still used today, e.g., for the calculation of the propulsion speed of swimmers with defined tangential velocity fields (known as squirmers). The Lorentz reciprocal theorem can also be used to relate the swimming speed of a micro-organism to the surface velocity which is prescribed

by deformations of the body shape via cilia or flagella. One of the flows for which the mathematical approach is relatively easy and which has been studied intensively from the 19th century is the flow of groundwater (see also § 6.1). This was possible as these flows were supposed to be laminar and the streamlines to be parallel to each other. Furthermore, boundary layers were (usually) not regarded in these flows.

Potential flows got the interest of fluid mechanics engineers again from the 1970s when the numerical simulation of flows started to become serious business. The available software and hardware were hardly capable of handling the Navier-Stokes equations and therefore the simulation of potential flows was taken up, even though they could not really be regarded as 'real' flows. In the 1980s Kees Vreugdenhil at the WL showed e.g., that the behaviour of waves (a free surface problem; see § 5.1.5) can, up to a certain point, be simulated by means of linearized equations and the theory of potential flows. The breaking of waves, however, remained out of reach with this approach.

## • 5.1.2 TURBULENT FLOWS AND TRANSITION

Most flows are turbulent: unsteady and chaotic, not repeating in detail. The turbulent state is opposed to the laminar state. The difference is significant, since the chaotic motions of turbulent flows produce much larger values for drag, and for mass and heat transfer, than corresponding laminar flows. Despite the omnipresence of turbulence and much research on the phenomenon, it can still be described as one of the last great unsolved problems of classical physics. There is no comprehensive theory of turbulence, although much partial qualitative understanding has been achieved.

### BURGERS

When Burgers started in Delft in 1918, turbulence was for the most part terra incognita. Burgers knew how Lorentz had approached turbulence some years earlier (see § 2.4) and was confident that he could contribute to the exploration of this intriguing phenomenon. To many scientists of those days turbulence still seemed a 'problem' which could be solved within a foreseeable amount of time.

It was his work on vortex motion in the early 1920s which stimulated Burgers to turn towards turbulence. In the *Annual Review of Fluid Mechanics* of 1975, he wrote:

"In 1923 I attempted to construct a theoretical model for turbulent flow between two parallel walls, in which an assumed distribution of shearing forces together with a distribution of viscous dissipation was introduced, based upon a kind of superposition of many of Lorentz's vortices. The model could be arranged in either of two ways: it could give a resistance proportional to the 1½-power of the mean

flow velocity ... or it could give a resistance proportional to the square of the flow velocity ... As Blasius' law for pipe and channel flow said that the resistance should be proportional to the 7/4-power of the velocity, an intermediate model would be needed. It looked as if this could be obtained by introducing some randomness in the arrangement of the vortices, but no appropriate solution was found."

In the meantime, Burgers shifted his attention to the results from Van der Hegge Zijnen's hot-wire measurements in the turbulent boundary layer (see § 6.2.1). However, his main interest in turbulence still concerned the theoretical approach. Like several of his foreign colleagues, he started to study statistical theories of turbulence, but soon discovered several difficulties:

"What stuck in me was an idea about the importance of the dissipation condition for turbulent channel flow: all the energy put into the system by the pressure difference driving the mean flow should be dissipated, for a (small) part in the viscous dissipation associated with the mean flow, and for the larger part in dissipation connected with the turbulent vortex system. For several years I played with the hypothesis that a statistical theory of turbulence might be built upon the example of the statistical theory used in the kinetic theory of gases or in other conservative systems, provided the condition of constant energy content was replaced by a dissipation condition. It became clear, however, that this method would lead to 'equipartition of dissipation' for all degrees of freedom of the system, and as there is an infinity of degrees of freedom so long as one keeps to the picture that the fluid is a continuum, there is the danger of infinite total dissipation."

The results of this 'playing' were presented in 1929 at a conference in Aachen. In the same year Burgers published the first three of a series of seven papers on the application of statistical mechanics to the theory of turbulent fluid motion. In 1933, the next four of this series were published. However, he was not satisfied with his results, as he remarked to the English physicist George Trubridge who in the 1930s wanted to write a thesis on 'Burgers' theory of turbulence'. Thanks to Trubridge's reviews published by the Physical Society of London and in Science Progress, Burgers' theory of turbulence gradually became better known. However, even though Trubridge and Burgers seriously discussed the problems surrounding the theory, and the German mathematician Blumenthal (see also § 6.1) gave algebraic support, around 1936 Burgers was still not satisfied with the results he had found.

Around this time, statistics started to dominate research in turbulence: a shift took place from the 'eddy' models developed by Taylor and Prandtl in the 1920s towards a statistical description. Burgers took yet another road in the 1930s. He decided to restrict his attention to model problems, with which the essential aspects of turbulence could be treat-

ed. This eventually led to his work on an equation which became known as the Burgers equation (see § 6.1).
After the War, and a long period of stagnation in his own research, Burgers was informed about the developments in the USA and Britain. He realized that his treatment of turbulence had taken a direction which differed strongly from others like Kolmogorov, Onsager, Von Weizsäcker, and Heisenberg. Nevertheless, he still felt convinced of the value of his work and was very much encouraged in this opinion when the famous mathematician John von Neumann visited him in 1949. Von Neumann had been sent by the US government to survey the research on turbulence at European centres of fluid mechanics. His report for the Office of Naval Research sketches a good view of the discussions on turbulence running at that time. One of Burgers' most promising results mentioned by Von Neumann, concerned a particular type of turbulence in which the vorticity is concentrated into vortex sheets.

Apart from his work on the Burgers equation, Burgers continued to work on the 'classical' statistical theory of turbulence. In 1953, he and Morton Mitchner (from Harvard University) did pioneering work on turbulent flow including a mean motion (i.e., a constant velocity gradient). The same year, he also published "some considerations on turbulent flow with shear", in which he tested "a simple expression for the relative frequency of patterns of turbulence of various scales at different distances from the wall in a turbulent boundary layer".

After Burgers' emigration in 1955, Broer and Hinze were about the only scientists in the Netherlands who could have continued Burgers' theoretical research. But they did not do so. Broer never published much about turbulence and Hinze was mainly working on his famous textbook (see § 3.3.2). In the 1960s and 1970s experiments were carried out under Hinze's guidance at the Laboratory of Aerodynamics and Hydrodynamics on the transition to turbulence in the boundary layer. During these experiments hot-wire measurements were performed (see § 6.2.1).

### LEIDEN, DELFT, TWENTE
In the 1970s many in the turbulence community turned towards research on coherent structures. In Delft the Japanese Ueda did experiments related to this, which led to Hinze's only paper in the Journal of Fluid Mechanics (in 1975), entitled 'Fine-structure turbulence in the wall region of a turbulent boundary layer'. Both Ooms (successor of Hinze), and Nieuwstadt (successor of Ooms), spoke about coherent structures in their inaugural lectures and research on these structures, both experimental and numerical, became one of the priorities under Nieuwstadt.

It was not only Delft where research into the fundamental aspects of turbulence was done in the first decade of this century. Around 2003 a group of physicists from Leiden and four other cities around the world discovered a new manner in which turbulence could arise. They had performed experiments on superfluid helium (a much-studied liquid in Leiden since Kamerlingh Onnes' research of the early 20th century) and discovered that the transition from laminar to turbulent flow was determined by the temperature. Below a critical temperature it appeared that the vortex lines in the helium were no longer stretched and 'quiet', but that they started to behave chaotically and to multiply. The very low temperatures and the special characteristics of superfluid helium enabled the researchers, as they said, to study turbulence in its purest form. Their results were published in Nature. Transition to turbulence has always remained a research theme in the Laboratory for Aerodynamics and Hydrodynamics, under Nieuwstadt and later under Jerry Westerweel. For the transition in flows of non-Newtonian fluids a 32 m long pipe flow facility was built in the 1990s. But much more publicity was gained in 2004, thanks to a publication in Science, by the results of an experiment done by PhD student Cas van Doorne, together with scientists from Germany, the UK and the USA. In Delft measurements took place in a pipe of 26 m length to find 'nonlinear travelling waves', a kind of coherent structures, which had been found by others from numerical simulations using the Navier-Stokes equations. These structures seemed to be unstable and many doubted the existence of them in real turbulent flows. The experiments showed that they did exist, and the measured velocity fields compared quite well with the numerical ones. In this experiment stereoscopic PIV was used, a technique which had largely been developed in the Delft laboratory (see § 6.2.1). "Our main contribution to the problem of turbulence is that we could show that principles from nonlinear systems theory appear to apply to this type of turbulent flow," postdoc Björn Hof told PhysicsWeb in 2004.

In 2005 another remarkable discovery was made in the same laboratory in Delft. Westerweel and others had studied the mechanics and transport processes at the bounding interface between the turbulent and nonturbulent regions of flow in a turbulent jet. Their results led them to the conclusion that there is small-scale eddying motion at the outward propagating interface (nibbling, as they called it) by which irrotational fluid becomes turbulent. The large-scale eddies, they concluded, are not the dominant eddies in the entrainment process, as was generally thought.

As the 'role' of coherent structures in turbulent flows became clearer, the idea arose that it might be possible to 'control' turbulence and get drag reduction by manipulating these structures. This initiated research on two drag reducing methods: modifying the geometry of the walls, and adding polymers to the liquid (the last had already been studied experimentally at the University of Amsterdam in

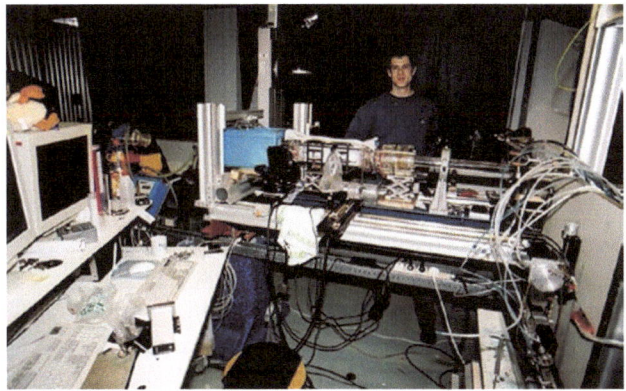

↑Cas van Doorne preparing the experimental facility in Delft with which the nonlinear travelling waves in turbulent flows were detected for the first time. For this stereoscopic PIV (see § 6.2.1) was used, with two 1000 Hz cameras. These were provided to the research team by the cameras' manufacturer; buying them was too expensive at the time. In the foreground on the right the facility with which laminar injections could be given to disturb the main flow. (courtesy of TU Delft/ Jerry Westerweel)

→Lohse at the Taylor-Couette facility at the University of Twente. In his Physica lecture of 2011 Lohse called the TC setup one of the two 'fruit flies' of fluid mechanics; the other one is the Rayleigh-Bénard convection cell (see § 5.1.7). With these 'fruit flies' new ideas and concepts can be tested in a rather uncomplicated manner. (courtesy of TU Eindhoven archives / photo by Bart van Overbeeke)

the 1970s). STW en FOM (see § 4.4) supported the research on drag reduction in Delft and Eindhoven from 1983. In Delft the effect of micro-grooves (or riblets) was investigated. So-called ejections and sweeps were identified since their role was considered important.

Around 2005 in Twente drag reduction by means of bubble injection was investigated in a Taylor-Couette facility. By injecting bubbles in the liquid between the rotating cylinders a strong drag reduction was found (up to 30%). But when the smooth surface was made rough with perspex strips, the reduction was much lower. Application of the bubble technique to ships therefore seemed to be not very effective since ship hulls are usually not very smooth.

In 2011 Lohse had the honour to give the Physica lecture, organized by the Dutch physics community. The main message of his lecture was: the image of turbulence which had been given by the Russian scientist Kolmogorov in the 1940s and which had been a 'paradigm' for many decades, had shown fractures in recent years. In Kolmogorov's image fully developed turbulence, at high Reynolds numbers, was homogeneous and isotropic. Numerical simulations seemed to confirm this image, but doubts had arisen as to whether this image was still valid for 'real turbulence', i.e., turbulence in real containers with boundary layers. Real turbulence

could switch to different states, and not just one. Among the experimental results which had led Lohse to this conclusion were those obtained by his group from studies on the flows in a Rayleigh-Bénard convection cell (see § 5.1.7).

## • 5.1.3 VORTICES AND ROTATING FLOWS

Many natural and technological flows are vortex-dominated. A vortex is a 'structure', e.g., a tube, of concentrated vorticity. Vortices exist on many scales: there are small ones in turbulent boundary layers (coherent structures), larger ones under the wings of airplanes. Still larger vortices are tornados and hurricanes, and then there are very large-scale vortices like the polar vortex or the Great Red Spot of Jupiter. The generation, interaction, and dispersal or mixing of vorticity plays a profound role in a wide class of applied, geophysical, and fundamental fluid flows, and this explains the longstanding research on vortical flows.

### BURGERS
In 1918 Burgers had already mentioned vortices in his inaugural speech, and gave the vortices related to the flight of birds as an example. The first experiments that were done in his laboratory in 1920–1921 were related to vortical flows (see § 6.2.1). In the early 1920s, Burgers' first three papers

appeared in which the movement of bodies in fluids and the related resistance was treated: one on the distribution of vorticity around bodies, one on the connection between generated vortices and resistance, and his first attempt to tackle turbulent flow.

Though his mathematical skills proved of great value, he encountered several difficulties which led him to a search for new methods of attack. His work on vortices directed his attention to the theory developed by the Swedish physicist Oseen (see § 6.1). In the Annual Review of Fluid Mechanics of 1975, Burgers remembered: "A publication of 1920, in which patterns of flow around a body were discussed as resulting from the interplay (or 'competition') between convection of vorticity by the mean flow on one hand and diffusion of vorticity on the other, had helped me to see the meaning of Oseen's theory of flow around a body with its unexpected sheets of discontinuity, as a special case of a more general problem. [My approach] took away the strangeness of Oseen's solution and gave it a place as an instance of a method of treatment with wider possibilities." Furthermore, Burgers could apply Oseen's theory to boundary layer theory: "I began to see a relation between certain aspects of Oseen's work and Prandtl's boundary layer theory, and I constructed an intermediate picture by making use of a transformation of the equations for two-dimensional flow, given by Boussinesq".

In a footnote to a paper of 1940, Burgers regarded an exact solution of the Navier-Stokes equation for which the vorticity can be written as (using cylindrical coordinates):

$$\omega = \frac{A\Gamma}{2\pi v} \exp\left(-\frac{Ar^2}{2v}\right)$$

where A is a constant, $\Gamma$ is the circulation of the vortex, and v is the viscosity. It still is one of the few exact solutions known and has since been called the 'Burgers vortex'. A remarkable feature of this structure is the fact that dissipation is independent of v. Today there is a renewed interest in this flow phenomenon, due to the discovery of the emergence of high-vorticity regions concentrated in tube-like structures in turbulent flows. The tubes are generally interpreted as vortex tubes which are stretched and concentrated, in a manner analogous to the Burgers vortex.

### EINDHOVEN

Since the late 1980s much research, both experimental and numerical, on vortices has been performed by Van Heijst (see § 4.1.2), Herman Clercx, and co-workers. He started this work when he was still at the Institute of Meteorology and Oceanography of the University of Utrecht and continued it at the TU Eindhoven. While in Utrecht Van Heijst became interested in ocean vortices.

One of the research projects was related to so-called 2D

turbulence. Whereas in 3D turbulence energy is transferred from larger to smaller structures and the larger structures disappear in time, in situations where vortices can only move around in a flat space, the situation is the other way around. In 2D there is self-organisation into larger structures and an inverse energy cascade (energy flux to larger scales). This is clear from the coherent vortex structures which are formed in 2D situations and which are persistent. Large-scale atmospheric and oceanic flows are to a first approximation 2D and therefore the study of 2D vortices is an interesting field. The Eindhoven team succeeded in creating 2D turbulence by arousing a pattern of dozens of 2D vortices in an electrolytic liquid by means of magnets and electric wires causing Lorentz forces.

In the early 1990s research was done on dipolar and tripolar vortices and on the behaviour of 2D coherent vortex structures in stratified fluids. Van Heijst and his co-workers showed for the first time, in 1989, how tripoles could be made from a cyclonic vortex in a rotating fluid. The results gained in Eindhoven also shed light on what happens when 2D vortices encounter walls and the consequences of these effects for flows in the vertical direction, e.g., the displacement of suspended material and material lying on the bottom of the basin.

### • 5.1.4 MIXING AND TRANSPORT PROCESSES

Burgers and his co-workers in Delft had never been involved in mixing and transport processes (transfer of mass, momentum, and heat) until in 1952 an engineer from India, Acharya, started to do experiments in the Laboratory on the 'momentum transfer and heat diffusion in the mixing of coaxial turbulent jets surrounded by a pipe'. Acharya was an aeronautical engineer who had somehow come into contact with Van der Hegge Zijnen. Probably Van der Hegge Zijnen had met the 'problem' of mixing coaxial jets after he had started to work for Shell (see § 3.3.2) and had suggested the subject to Acharya. The Indian would be Burgers' last PhD student in the Netherlands.

As for heat transfer, Van der Hegge Zijnen had been doing experiments on the spreading of heat from a hot-wire in a turbulent shear flow around 1950, in his laboratory at Shell. In 1951 he and Hinze had presented a paper about this research at a conference in London. In 1948 they had already presented results on the 'transfer of heat and matter in the turbulent mixing zone of an axially symmetric jet' at the 7th International Congress for Applied Mechanics, also in London.

For many years, Shell was the company in the Netherlands which stimulated research into transport processes. As we have seen in § 3.3.2 Shell, or actually BPM, had an important role in the start of this research at the TH Delft. For more than a decade it was professor Kramers who would lead this research and would become well-known as an expert in

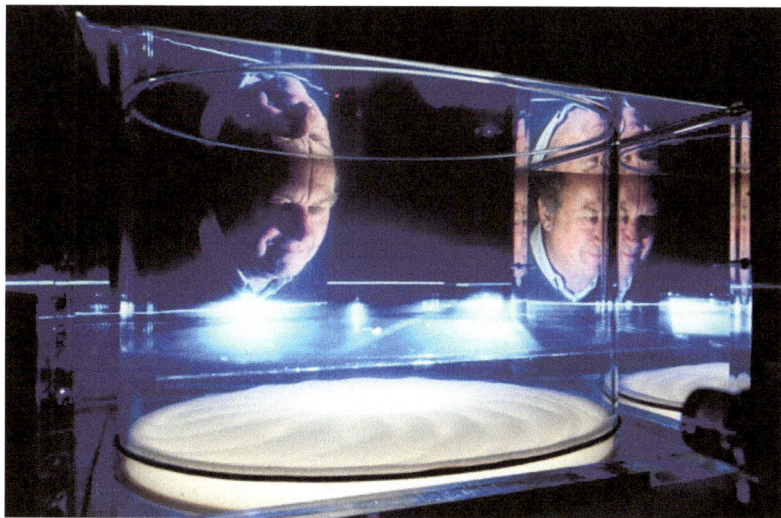

↑Van Heijst observes the pattern formation in a sediment layer, driven by a large vortex flow in his laboratory in Eindhoven. (courtesy of TU Eindhoven archives / photo by Bart van Overbeeke)

→The Earth's rotation has an influence of many a flow phenomenon in nature. In the WL in Delft model experiments were run on large water areas for which this rotation could not be neglected. To scale down the rotation in a simple but effective manner, rotating rods were placed in the water. These rods were introduced by WL director Schoemaker in the 1950s; before, one had to use large turning tables to simulate the Earth's rotation. This photo was taken during a visit of the King and Queen of Thailand to the WL in 1960. (courtesy of TU Delft / photo by Fotografishe Dienst via Beeldarchief / CC BY)

↑The Van Heijst group in Eindhoven became famous for the attractive visualisations of two merging, and later separating, counter-rotating monopolar vortices. The most common coherent structures are the axisymmetric (monopolar) vortex, with circular streamlines, and the vortex dipole, both of which have been found to arise in a variety of situations under different forcing conditions. The photo on this calendar shows the result of a head-on collision of two dipolar vortices in a stratified fluid environment. The original vortices have exchanged a partner to form two new dipoles. This result was published in *Nature* in 1989 and reached the Nature calendar for 1990, as can still be seen in Van Heijst's office. (courtesy of TU Eindhoven / GertJan van Heijst)

↑The experimental set-up as used by Acharya in Delft and shown in his thesis: two jets are produced at opposite locations and they meet in the big drum in the middle. (courtesy of Burgers Archives / TU Delft)

↑At the end of the 1940s Hinze and Van der Hegge Zijnen were both working at Shell's laboratory in Delft/Rijswijk. One of their joint projects concerned the transfer of heat and matter in the turbulent mixing zone of an axially symmetric jet. This photo of their experimental facility was published in their paper in the very first volume of *Applied Scientific Research A* (see § 6.1). (courtesy of Springer Nature)

## ROBERT B. BIRD ON HANS KRAMERS

Hendrik ('Hans') Kramers was to play an important role in the development of fluid dynamics in the Netherlands. He was the son of the Islamic scholar J. H. Kramers and the nephew of the famous theoretical physicist H. A. Kramers, both professors at the University of Leiden. From 1934 to 1941 he was a student at what would later be called the Technical University of Delft. After working for several years at the TNO (Organization for Applied Scientific Research) in Delft, and at Royal Dutch Shell in Amsterdam, in 1948 he was appointed at the very young age of 30 to a professorship in Delft to develop the field of engineering physics, with emphasis on fluid dynamics, transport phenomena, chemical engineering kinetics, and process control.

Fluid dynamics, heat transfer, and mass transfer are all subjects which were developed in Europe. However, these subjects are often intertwined: it is very seldom that one of these occurs entirely alone. Hans Kramers was possibly the first professor in Europe to recognize that the three topics above should be taught in a single course, so that students could take advantage of the similarities (and differences) between the three. The mathematics used in the three areas is very similar. In 1956, Hans turned out a set of mimeographed notes entitled Physische Transportverschijnselen to use in teaching the subject of transport phenomena to engineering students.

At the same time, he and his students were pursuing experimental studies of subjects, primarily in the area of mass transfer and diffusion:

- S. Lynn, J. R. Straatemeier, and H. Kramers, Chem. Eng. Sci., 4, 49-67 (1955): diffusion into a falling liquid film (gas absorption);
- H. Groothuis and H. Kramers, Chem. Eng. Sci. 4, 17-25 (1955): mass-transfer rates during drop formation at a capillary tip;
- H. Kramers and P. J. Kreyger, Chem. Eng. Sci., 6, 42-48 (1956): diffusion into a falling liquid film (solid dissolution).

Hans had very close relations with Professor P. V. Danckwerts at Cambridge University in England. These two, although very different in personality, exchanged many ideas about mass transfer and diffusion through the years. Perhaps this interaction influenced some of the experimental set-ups in Hans's laboratory.

Since I had been teaching transport phenomena to graduate students since 1953 and began work on a course for undergraduates in 1957 (see Recent Advances in the Engineering Sciences, McGraw-Hill, New York (1958), pp. 155–177), I decided to apply for a Fulbright lectureship, and take a sabbatical to spend the spring of 1958 at Kramers' laboratory in Delft. He invited me to give a set of lectures during that period. That was very easy for me to do, inasmuch as W. E. Stewart and E. N. Lightfoot had joined me in preparing a first draft of what would become our textbook entitled Transport Phenomena.

I quite enjoyed interacting with Hans. He kindly offered me a room in his house, and we would walk together from his home to the lab every day and discuss many topics of mutual interest. I was intrigued by the many laboratory experiments that he had his students working on, particularly in the field of mass transfer. I was also interested in noticing the friendly way that he interacted with his students, and how much they respected him. Although he could be quite strict and critical, he always did this with a smile and a twinkle in his eye. He guided 164 students to the 'Ir.' (Engineer) degree, and 15 to the doctorate, of whom 10 went on to become professors themselves. Many of his students became prominent in research, engineering, and management.

There is no question that Hans's unification of the field of transport phenomena played an important role in the advancement of the design and operation of the industrial processes in the field of chemical technology. Of equal importance is his work in the field of fluid dynamics with chemical reactions. This work was summarized in the monograph Elements of Chemical Reactor Design and Operation by H. Kramers and K. R. Westerterp, Netherlands University Press (1968).

Hans' accomplishments have been recognized by election to the Royal Netherlands Academy of Sciences, an honorary doctorate from the Eindhoven University, honorary foreign member of the National Academy of Engineering (USA), a Knight of the Order of the Lion (Netherlands), and many others.

↑On July 2, 1963 Kramers gave his farewell lecture at the TH Delft. He told his audience that it was important to keep the training of students in research skills at a sufficiently high level. Notice the elementary mixing process drawn by Kramers on the chalkboard. (courtesy of TU Delft / photo by Fotografishe Dienst TU Delft via Beeldarchief / CC BY)

←In the large basin in the Stevin III Laboratory – to-day called Waterlab – of the Department of Civil Engineering in Delft waves could be studied. This photo was taken around 2005, when Battjes had already retired. Some five years later the basin was removed for financial reasons. (courtesy of Delft University of Technology)

the field which has become known as transport phenomena ('fysische transportverschijnselen' in Dutch).

Transfer processes in boundary layers were also studied at Shell, as is evidenced by a paper by Merk, published in the Journal of Fluid Mechanics in 1959. As we have seen in § 4.1.1 Merk had already been working on heat and mass transfer before he came to work with Shell. While professor in Delft from the early 1960s Merk turned to the field of rheology. From about 1970 he also started to lecture on and write about TIP: the thermodynamics of irreversible processes. Together with Gerard Kuiken, Merk has given a clear derivation of the Maxwell–Stefan diffusion equations and he has applied the TIP theory to rheology.

## • 5.1.5 FREE-SURFACE PHENOMENA

Two phenomena involving a free surface (a surface with zero parallel shear stress) have a long tradition in the Netherlands: surface water waves and droplets. The first topic was, and still is, a field of interest for engineers and scientists working in hydraulics and the branch of fluid mechanics related to hydraulics. The interest in the second topic started mainly in the industrial world of process technology.

### WAVES

Burgers never worked on either of these phenomena. It seems that only at the KNMI some interest in the theory of waves existed, at least from the 1940s. Timman got interested in waves due to his work on ship hydrodynamics (see § 6.1). Broer, who had been Burgers' colleague for some years (see § 4.1.1), started with the mathematical approach to waves when he was still in Delft (with a paper entitled 'On

the propagation of energy in linear conservative waves'), but only became seriously involved in wave theory (which was not only applicable to water waves) after he had moved to Eindhoven in the 1960s. One of his PhD students there was Brenny van Groesen (1949) who became professor at the University of Twente. Van Groesen became a colleague of Zandbergen (see § 4.1.3) and the group called Applied Analysis & Mathematical Physics became known partly for its work on waves and their numerical simulation. One of the activities of this group has been a collaboration with researchers in hydraulics in Bandung, Indonesia.
It seems that the serious study of water waves only got its momentum in the 1970s, with the appointment of Battjes in Delft (see § 4.1.1) and a growing interest in waves at the WL. One of Battjes' achievements has been a mathematical model for the dissipation of energy during the breaking of (irregular) waves. Battjes' theoretical predictions were compared to the results of experiments performed by a student named Hans Janssen and therefore the model has become known as the Battjes–Janssen model. It became famous around the world and has since been used by wave researchers. Later, the model was integrated in a more comprehensive wave simulation program. The making of the program was financed by the US Navy, which insisted that it would become open source. The model became known as SWAN; it has been used worldwide and is now in its third 'generation'.

Sloshing is a phenomenon related to wave behaviour. To study such a phenomenon without the hindrance of gravity forces, an environment of so-called microgravity is needed. One of the possibilities of creating such an environment is the performance of experiments aboard a space vehicle. The

NLR became involved in the numerical simulation of sloshing from the late 1970s when such an experiment was carried out aboard a satellite. But numerical techniques were still limited at that time and only simple 2D models of liquid motion under microgravity could be studied. A major boost took place in 1995 with the development by the FLEVO (Facility for Liquid Experimentation and Verification in Orbit) of the Sloshsat which was launched in 2005. It was a mini-satellite designed and built by the NLR to experimentally study liquid dynamics in a water tank onboard the spacecraft. Numerical researchers at the University of Groningen started simulations of the sloshing phenomenon. The development of their simulation method for fully 3D free-surface flow became known as ComFlow (see also § 6.3).

## DROPLETS

When Hinze still worked for Shell in the 1950s, he derived an expression to predict the maximum size of droplets in a turbulent flow. In the formula, published in 1955, one can find a model constant, the interfacial tension at the droplet surface, the density of the carrier fluid, and the so-called turbulence dissipation rate. This correlation is still used in engineering design codes to determine, for example, the efficiency of separation processes.

For a long time it was impossible to compare (numerical) models of droplets with experimental observations. Only with the arrival of sophisticated high-speed cameras could progress be made. It was Lohse in Twente who was one of the first to acquire such cameras and to build a facility in which moving images of the deformation of droplets could be made. One of the well-known images from Twente shows the 'splashing' of a droplet which enters a liquid surface.

## • 5.1.6 COMPRESSIBLE FLOWS

Compressible flows are those in which the changes in pressure from place to place in the flow are so large that the density of the fluid is changed. These flows present special difficulties: shock waves can arise which move faster than the speed of sound, and temperatures can be high and non-uniform, causing a number of effects that are difficult to predict. Compressible flows are well-known from aeronautics but there are many other areas where they occur, such as in vacuum technology and certain manufacturing processes.

## ACADEMIC RESEARCH

During the 1920s and 1930s, Burgers' interest was completely restricted to incompressible flow. Though the interest of fluid dynamics scientists in compressible flows had been stimulated during the 1930s by, e.g., some papers of G.I. Taylor, it was only during the Second World War that this field of research really got full attention due to the development of (supersonic) war aircraft and missiles.

Characteristically, Burgers' approach to a field of research unknown to him was to start his first paper (1943) with a relatively simple (i.e., one-dimensional) problem. In papers of the late 1940s, Burgers continued the study of shock waves and their interaction. He also hoped to establish a new experimental direction for his Laboratory with shock tubes. Shortly before the Second World War, agreements had been signed with the NLL: Burgers' laboratory would study supersonic flow, while the NLL would restrict its attention to subsonic flows. However, after the War the NLL also took up supersonic aerodynamics (see § 4.2.4 and 6.2.2). Furthermore, the TH Delft decided that work in this area had to be concentrated in the new Department of Aeronautical Engineering. For Burgers, this must have been another frustration which strengthened him in his decision to leave Delft.

During the same period, Burgers also got interested in gas dynamics, a subject which brought him somewhat back to his origin as a theoretical physicist. This interest was stimulated by his acquaintance with problems related to cosmic or astrophysical fluid mechanics. In the 1940s the Dutch Astronomers Club invited him to give a lecture on the borderlands between fluid mechanics and astronomy and there he met the famous Dutch astronomer Jan Oort. Stimulated by Oort, who supplied the astrophysical data, Burgers started working on turbulence in rotating interstellar gas masses. After his emigration to the USA in 1955, Burgers began to specialize in hypersonic aerodynamics. He started a seminar on high-speed and high temperature flow problems and studied the relation between the Boltzmann equation for gaseous systems and the equations of fluid dynamics. In 1969, aged 74, his book Flow equations for composite gases appeared.

From about 1950 Burgers' colleague Broer had also started the study of shock waves. In the late 1950s the Laboratory for Aerodynamics and Hydrodynamics did indeed own a small shock tube in which shock waves could be studied. One of the topics on which Broer wrote was acoustic relaxation, a phenomenon that was studied (also experimentally) from about 1970 in the group of professor Vossers in Eindhoven, by Rini van Dongen and others. Leen Noordzij in Twente studied shock waves in mixtures of liquids and bubbles under the guidance of Van Wijngaarden.

## SUPERCRITICAL WINGS

During the period after the Second World War the aeronautical world was confronted with the so-called 'sound barrier', the rapid drag increase at high transonic speeds (i.e., just under the speed of sound) and the occurrence of shock waves in the flow around wings. These shock waves were caused by the supersonic flow over the wings. The formation of shock waves was delayed with increasing speed by changing the sweep angle of the wings. Highly swept-back wings, however, had aerodynamic and structural disadvantages. Vague ideas arose

↑Lohse with a photo showing results of one of his experiments on droplets. On the wall a splashing droplet can be seen. (courtesy of Océ / photo by Gijs van Ouwerkerk)

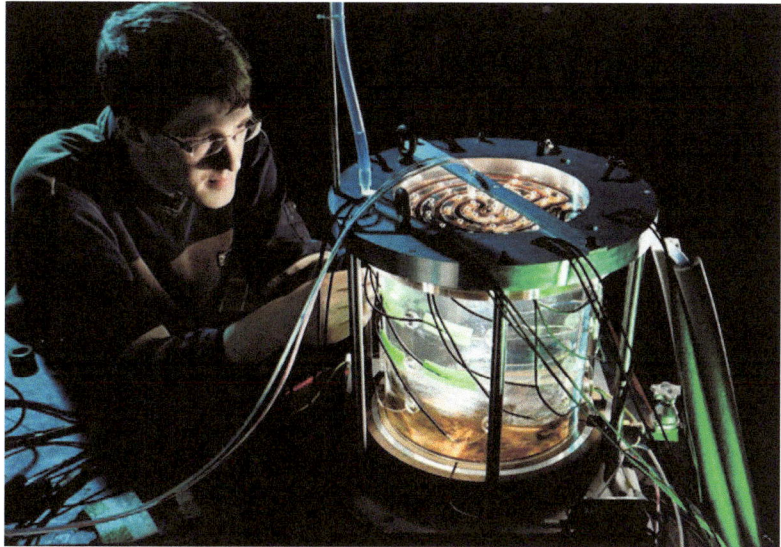

↑Richard Stevens was one of Lohse's co-workers during the research on the Rayleigh-Bénard convection cell. Their experiments were (partly) performed in the USA, but here Stevens is seen with an RB cell of the TU Eindhoven. (courtesy of TU Eindhoven archives / photo by Bart van Overbeeke)

←The first supercritical wing designed by the NLR was the result of a collaborative study with Fokker. Here the wings are tested in the High-Speed Wind Tunnel of the NLR in Amsterdam. (courtesy of Stichting Behoud Erfgoed NLR)

about other means of delaying the occurrence of shock waves. Around 1960 the NLR started, as the first institute in the world, to develop shock-free aerofoils along theoretical lines, that is aerofoils with local supersonic flow on which no shock waves would appear when the local supersonic flow decelerated to subsonic flow. Several scientists at the NLR started to work on the problem. In 1967 they could show that it was theoretically possible to design shock-free transonic aerofoils and that these flows were stable. Thereafter methods were developed for the design of real 3D wings. These activities were carried out in the late 1960s when the digital computer gradually became of age and when it became possible to compute the aerodynamic forces on complete wings and aircraft configurations. In the same period wind tunnel experiments were performed which showed that the new aerofoil design was indeed shockwave-free. From 1969 feasibility studies were carried out in cooperation with Fokker in order to arrive at wing designs which could be incorporated in a civil transport aircraft. During the period 1975–1977 several wing-body configurations were designed and tested in one of the NLR wind tunnels. The final result of all the years of research (and millions of guilders) on the supercritical wing was the construction of the successful Fokker F100 in 1983–1984.

## •5.1.7 BUOYANCY DRIVEN FLOWS

The density of flowing material can differ from place to place, because of temperature variations or because the composition is not uniform. Where the density is lower, the material tends to rise, where it is higher, the material tends to sink. A flow produced by these effects is called buoyant convection (a term which doesn't have a really satisfying translation in Dutch, by the way). Buoyancy driven flows can occur in heated rooms, around fires, in energy storage systems, inside the planet Earth and in atmospheric and oceanic systems.

Many geophysical and astronomical phenomena are driven by highly turbulent fluid dynamics. These dynamical flows are of-

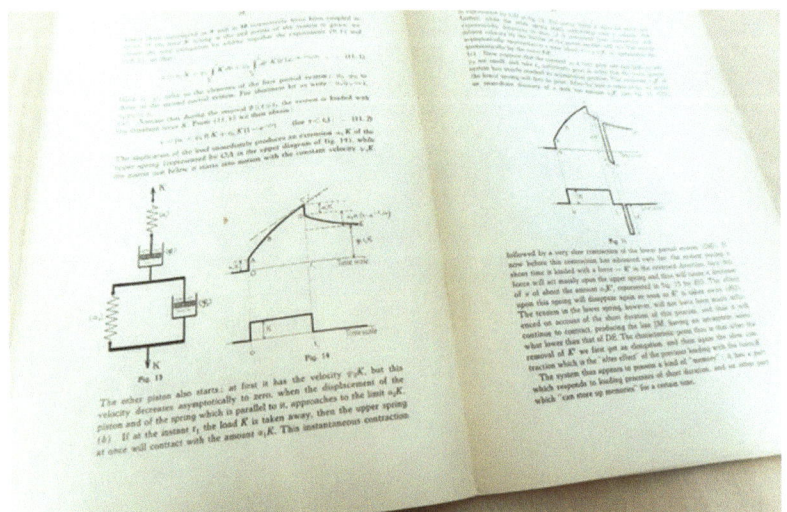

↑In the *First Report on Viscosity and Plasticity* (1935) Burgers introduced a combination of already existing models for visco-elastic materials. This new model has been called after Burgers for some time but today it seems to have become obsolete. Burgers stressed the fact that this model had "a kind of 'memory'". (courtesy of Burgers Archives / Delft University of Technology)

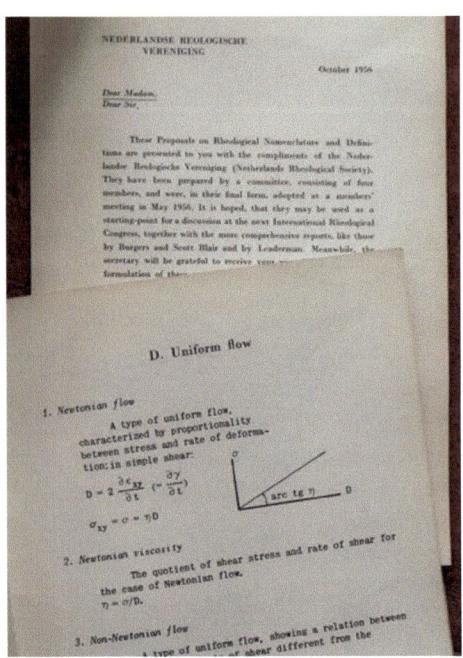

↑The Dutch Rheological Society was founded in 1951 and continued the work of the Viscosity Committee with regard to nomenclature. This booklet was published in 1956. (courtesy of Ned. Reologische Ver.)

ten driven by the buoyant rising and falling of fluids of different densities and strongly affected by the rotation of the celestial body through Coriolis forces.

In the Netherlands, the amount of research projects on heat transfer by convection had been small as compared to other fields of fluid mechanics. De Vries (see § 4.1.2) had been working on this theme from the late 1950s, and Hoogendoorn (see § 4.1.1) started a new group which would study convection, by means of experiments and numerical simulations, in the early 1970s. It is not surprising that both professors took the initiative, in 1971, to start a Contact Group for Fundamental Heat Transport.

The instrument to study convection has been, for over a century, the so-called Rayleigh-Bénard convection cell: a simple container filled with liquid or gas, heated at the bottom and cooled at the top. In the 1990s most researchers considered the basic theory of the convection in this cell as basically solved. That is, the relationship between the so-called Nusselt number (an indication for the heat transport) and the Rayleigh number (an indication for the temperature difference) was known and unambiguous. But around 2000 Lohse (University of Twente) and others came with a new theory. The main idea of this was to decompose the energy dissipation rate and the thermal dissipation rate into their boundary layer and bulk contributions. From experiments and numerical simulations, also on rotating Rayleigh-Bénard cells, they discovered that the relationship between Nu and Ra was not unambiguous and apparently, different turbulent regimes could occur inside the cell.

# 5.2 FLOWS OF NON-NEWTONIAN MATERIALS (RHEOLOGY)

Though it must have been clear long before the 1920s that the flow behaviour of liquids like blood and paint (which are now called non-Newtonian) was different from that of water, it was only then that the first publications appeared and the name 'rheology' was introduced (which would finally become 'reologie' in Dutch, without the 'h').

The Dutch Rheological Society was founded in 1951, by the Royal Netherlands Academy of Arts and Sciences. Before this Dutch scientists had already been involved in research, mainly theoretical, which we would now indicate as 'rheological' but this term was only used from the 1950s. Today the Society is a member of the Bond voor Materialenkennis (Alliance for Knowledge of Materials), which makes clear that rheology is not just a branch of fluid mechanics. Most of all rheology is about the characteristics of non-Newtonian materials which include an important number of interesting liquids such as molten polymers.

## • 5.2.1 EARLY DEVELOPMENTS IN DELFT

The area of plastic deformation had become attractive to many of the scientists working in fluid mechanics from the 1920s (e.g., Von Kármán, Taylor, and Prandtl). Burgers got an opportunity to become engaged in research in this area when in 1932-1933 the KNAW (at the instigation of Kruyt) installed the Viscosity Committee, consisting of researchers from several fields: physics, mechanics, chemistry (especially colloid chemistry), and biology. Its purposes were rather ambitious: to gather information regarding the phenomena of viscous and plastic deformation, so as they present themselves in various domains of physics, chemistry, technology, and biology; to investigate the relations existing between these phenomena; to make proposals for a nomenclature which should obviate existing uncertainties in the various domains; and to study the methods used for measurements of viscosity and of related properties of matter. Burgers, as the representative for mechanics, became the secretary of the Committee. During the 1930s, two impressive reports were published, i.e. the first and the second Report on Viscosity and Plasticity.

Viscoelastic materials can be modelled by representing the molecules as a combination of viscous dampers and elastic springs. In the First Report on Viscosity and Plasticity (1935), Burgers introduced a rheological model combining the older Maxwell and Kelvin-Voigt models. In 1956, the term 'Burgers element' was proposed by the Nomenclature Committee of the Dutch Rheological Society for this model, though today usually the term 'Burgers body' of 'Burgers fluid' is used. This model appeared to describe the response of materials such as asphalt and concrete quite well and has also enjoyed a good deal of popularity in the field of geomechanics.

In the post-war period, Burgers remained active in rheology. He became editor-in-chief of the Monographs on Rheology, founded in 1948 and published by publishing house North Holland. Furthermore, he was active, with Houwink of TNO and others, in the organisation of the First International Congress on Rheology, held in 1948 in Scheveningen (near The Hague). The Dutch contributions to this congress show the broad range of topics which were discussed: two engineers from the State Mines on the measuring of viscosity of settling suspensions; Jan (J.J.) Hermans (who became an expert in polymers) on swelling; two engineers from the Rubber Institute of TNO on the rheological properties of rubber; a researcher from the Algemeene Kunstzijde Unie (later AKZO) on the optical relaxation times in cellulose solutions; and a researcher from the Institute for Graphical Technology of TNO on 'typographic inks'.

Burgers also became secretary of the Joint Committee on Rheology of the International Council of Scientific Unions. For this Committee he wrote a paper, with the British rheologist Scott Blair, on rheological nomenclature. In 1964, when his interest in rheology had already diminished away, he was awarded the Bingham Medal of the Society of Rheology. Of Burgers' work in rheology published after 1940, we mention one paper which had actually been a kind of occasional

project. The Dutch colloid-chemist Hendrik Bungenberg de Jong (who had contributed to the first report of the Viscosity Committee) had performed experiments on the oscillatory movements presented by certain soap solutions in spherical vessels and asked Burgers for a theoretical treatment. This paper again shows Burgers' remarkable ability to handle a seemingly complex problem by using a range of mathematical tools, leading to a useful (instead of general) solution of the problem.

In the 1950s research on rheological topics was still scarce in the Netherlands. One of the first PhD theses in which rheology played an important role was published in Delft in 1951. It concerned a 'colloid chemical and rheological study' of drilling fluids. The research was partly done at KSLA and financially supported by the BPM. Burgers himself was one of the two supervisors (the other was a chemistry professor) of the research leading, in 1954, to the PhD thesis entitled Investigations on the rheological properties of clay (in Dutch).

Professor Broer, whose publications were on a broad range of topics when he held the chair in Delft, also wrote on the flow of visco-elastic fluids (see also § 6.1). But it was Broer's successor, Merk, who would give rheology and non-Newtonian fluids a more lasting status in the Laboratory of Aerodynamics and Hydrodynamics; his inaugural lecture of 1962 was about rheological models. In the 1960s interest in visco-elastic fluids also started to grow among the chemists in Delft (see § 5.2.3). In the Kramers Laboratory, for example, experiments on polymer melts were started.

## •5.2.2 MORE RECENT DEVELOPMENTS

An activity somewhat related to rheology and with a long tradition is rheometry: measuring the properties (such as viscosity) of non-Newtonian materials. For these measurements special instruments have been used and developed, called rheometers. In the Netherlands rheometers have been developed at diverse places: TNO (during the 1960s), Unilever (see also § 4.3), the Laboratory of Physical Colloid Chemistry of the University of Wageningen (see also § 4.1), etc.

Colloid research has a long tradition with the University of Utrecht (see § 4.1.4) and Wageningen. Another place where colloids have been studied, was the University of Twente. The group of Mellema (see § 4.1.3) studied colloidal hard spheres dispersions and in the 1980s he, in cooperation with the colloid researchers in Utrecht, discovered their visco-elastic behaviour.

In his inaugural lecture of 1992 Mellema made a number of observations on the situation with regard to rheology in The Netherlands at that time:

• rheology had for a long time been a field of research which was almost exclusively populated by chemists

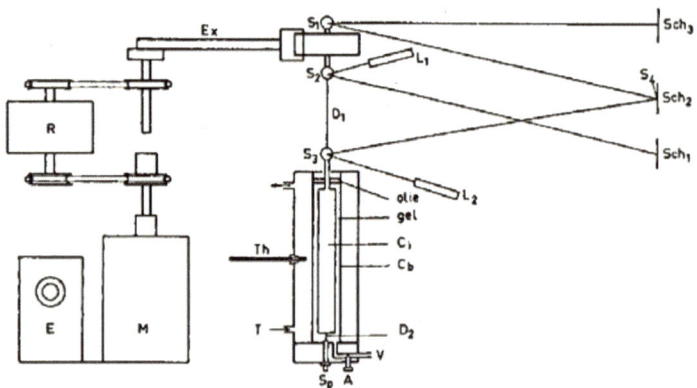

↑This rheometer was developed by TNO and used by several researchers, e.g., Herman Beltman for his PhD work related to "the influence of polymers on the rheology of aqueous systems" (Wageningen 1975). In the colloid group in Wageningen these kinds of measurements were usually related to research on foodstuffs. M is the engine which drives a cylinder Ci. The fluid is introduced between Ci and the cylinder Cb. Ci oscillates at various amplitudes. By means of lamps M and mirrors S light beams are pointed towards the scales Sch. The values read on these scales can be used to calculate the viscosity. (from: Beltman (1975))

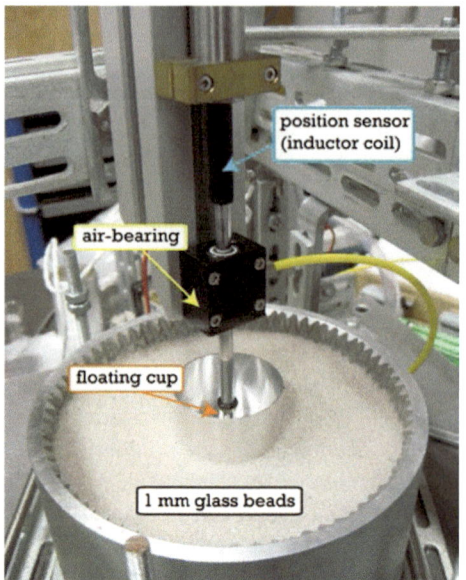

↑In the Leiden laboratory of Van Hecke experimental facilities like this have been used to study the behaviour of granular materials. If the material was stirred locally, it started to behave in a liquid-like manner and an effective viscosity could be determined for the system. It turned out that the granular medium used here is non-Newtonian. (from: Nichol (2011) / courtesy of Kiri Nichol)

but physicists were starting to become involved in this research also;
- there was still little collaboration between theorists and experimenters (which was also valid for fields outside rheology);
- rheology was finally getting recognition in the world of fluid mechanics.

This growing recognition was also due to work done in Delft and Eindhoven. In Delft Martien Hulsen started to develop numerical methods for flows of visco-elastic fluids from about 1985, under the supervision of professor Wesseling of the Department of Applied Mathematics but actually working in the Laboratory of Aerodynamics and Hydrodynamics. In his PhD thesis (1988) he showed that there was still a long way to go before numerical methods would be fast and reliable. In the 1990s Hulsen and colleagues started simulations of viscoelastic fluid flows using Brownian configuration fields, after it had become clear that a macroscopic approach based on the constitutive equations had serious restrictions. Only from the mid 1990s computer technology allowed a microscopic approach of the simulations. After Merk's retirement it took several years before the chair of Rheology was occupied again: in 1994 Ben van den Brule started a part-time professorship in Delft. Van den Brule, who had been a student of Van Wijngaarden in Twente, had worked on polymers at the Philips company and later worked for Shell where polymers

are studied which can be helpful in enhanced oil recovery. A rather new field of research related to rheology was that of granular materials, particles which are much larger than the particles of colloid suspensions. In 2007 FOM initiated a research program named Rheophysics. Goal of this program was a better understanding of the behaviour of so-called 'yield stress fluids': fluids (like shaving foam) which can show the behaviour of both a solid and a liquid. The transition from a liquid phase to a solid phase is called jamming and this phenomenon was investigated by professor Martin van Hecke and co-workers at the University of Leiden.

Several surprising phenomena related to non-Newtonian fluids have puzzled researchers for decades. As for so-called shear-thinning fluids this was the case with the Kaye effect. This effect occurs when a thin stream of a solution of such a fluid is poured into a dish of the fluid. As pouring proceeds, a small stream of liquid occasionally leaps upward from the heap. Michel Versluis of the University of Twente and co-workers showed that the Kaye effect works for many common fluids, including shampoos and liquid soaps. Around 2006 they revealed its physical mechanism through high-speed imaging. Their measurements could be interpreted with a simple theoretical model including only the shear thinning behaviour of the liquid; elastic properties of the liquid appeared to play no role. One of their videos of the Physics

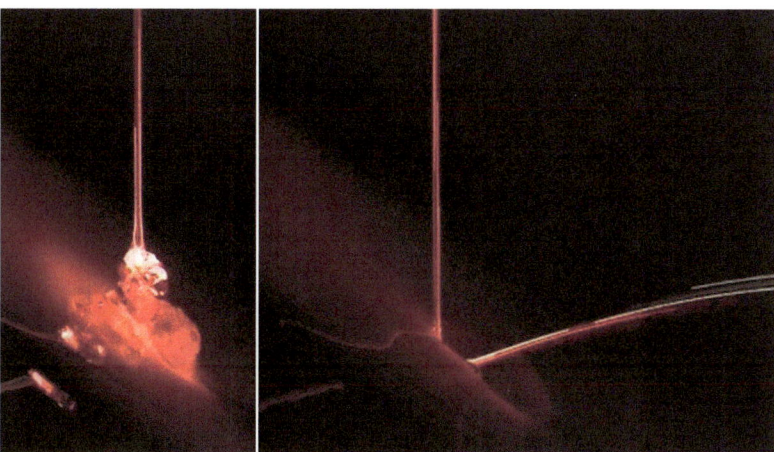

↑Shots from a video made with a high-speed camera of the Kaye effect at the University of Twente around 2006. (courtesy of Twente University / Michel Versluis)

↑This so-called multipass rheometer has been made and is used by the Polymer Technology group of the TU Eindhoven. This meter has a special slit (6 mm x 1.5 mm) which makes it possible to perform rheological and X-ray measurements at the same time. Since a very strong X-ray beam is necessary, the researchers have to take this instrument to European Synchrotron Radiation Facilities in Grenoble (France). With this rheometer the crystallization of polymers during their flow can be followed. (courtesy of Polymer Technology, Mechanical Engineering, TU Eindhoven)

of Fluids group of the UT was awarded, not for the first time, in the Gallery of Fluid Motion competition, organized by the *Physics of Fluids* journal.

## •5.2.3 POLYMERS

Research on the behaviour of molten non-Newtonian materials in production machines like extruders became very important after polymers had become widely used in both daily life and industry (e.g. as catalysts). But already during the Second World War Burgers got involved in a problem which can be regarded as closely related. An industrial company from Arnhem asked him to give suggestions for the improvement of the production of fibres from molten silicate for the production of glass wool. These fibres were produced by spraying the molten glass by means of steam jets. Burgers took up the problem, wrote a report but had to conclude that the changing viscosity of the silicate (due to a changing temperature) and other circumstances led to a flow problem which could not be solved with the existing theories. As for the extrusion and molding process with polymers, finding a theory was even more challenging. The behaviour of molten polymers is hard to predict since they possess a 'memory': in such a liquid the momentary stress is determined by the complete deformation history the fluid has experienced. Besides, the interaction of the melt with the

walls along which it is flowing, e.g., in extrusion processes, has appeared to be determining and can result in instabilities of the extruded fibres.

In the Netherlands, basic knowledge of the mechanical and thermodynamic behaviour of amorphous polymers was established mainly at the Plastics Institute of TNO during the 1950s and 1960s. Roelof Houwink (1897–1988) became a well-known polymer scientist at TNO. He had been educated in, had worked at Phillips since 1925 and from 1939 was director-general of the TNO's Rubber Institute in Delft. At the suggestion of Henri Brinkman (of the Brinkman number; see the introduction of § 6.2) a working group for fundamental extrusion research was established at the Central Laboratory of TNO in the 1960s.

One of the TNO researchers was Hermann Janeschitz-Kriegl (1924-2018), from Austria. In 1969 he became professor of 'Physics and Chemistry of Macromolecular Materials' in Delft and he has been called – by professor Frits Dijksman (Twente) - one of the founding fathers of 'industrial rheology as a science in the Netherlands and elsewhere'. Janeschitz-Kriegl was the promotor when Robert B. Bird (see § 5.1.4) became doctor honoris causa in Delft in 1977 "for his excellent merits in the field of physical transport phenomena in general and of the rheology of macromolecular liquids in particular". In Twente professor Van der Wallen Mijnlieff (see § 4.1.3) did research on macromolecular systems, including polymer solutions, but seems to have published little.

At the University of Eindhoven rheology really got a boost in the 1990s from the Polymer Technology group. Although research on polymers had been carried out in Eindhoven from the early 1970s, an important upscaling of rheology-related research took place after 'computational rheology' reached a higher level in the 1990s and 2000s (Hulsen moved from Delft to Eindhoven in 2001). Eindhoven became the 'secretary' of the Dutch 'onderzoekschool' (research school) for polymer technology which was founded in 1994 and which became Eindhoven Polymer Laboratories some ten years later.

From the 1990s Dutch researchers developed two new simulation techniques with which a microscopic modelling of the polymers can be combined with the simulation of a complex flow in macroscopic domains. The first one is the so-called Brownian configuration field method (already mentioned in § 5.2.2) and the other the so-called deformation field approach. From about 2005 numerical simulation of non-Newtonian liquids reached the 'adult' phase. In 2016 polymer researchers from Eindhoven were the first to perform 3D direct numerical simulations of the alignment of two and three rigid, non-Brownian particles in a viscoelastic shear flow.

In the 1990s polymers also got attention from the fluid mechanics community for a different reason. As has already been remarked in § 5.1.2 research was started into the influence of polymers on drag reduction in turbulent flows. Jaap den Toonder (who became professor of Microsystems in Eindhoven) did experimental and numerical studies on turbulent pipe flows and concluded in his PhD thesis of 1995 that "the key property for drag reduction by polymer additives is the purely viscous anisotropic stress introduced by the extended polymers, while elasticity has an adverse effect on the drag reduction".

Around 2003 Daniel Bonn, then in Paris and later professor at the University of Amsterdam, discovered that so-called melt fracture ('smeltbreuk' in Dutch) is inherent to the flow of molten polymers in the tube during the extrusion process. Up to that time, engineers had supposed that fracture was due to circumstances at the entrance of the tube or at the exit of the tube. Research in Leiden and Paris showed that above a certain critical value of the ratio of elastic and viscous forces, turbulence would develop inside the molten polymer in the tube.

# 5.3 TWO-PHASE FLOWS

Though many single-phase flows still give researchers enough work (and headaches), some researchers have also turned their attention to flows which are usually even more complicated: flows in which liquids, gases and solid particles come together in a huge number of variations. The famous British fluid mechanicist George Batchelor sketched the situation as it was in 1971: "As a newcomer to the field [i.e., two-phase flows], I can take the position of the small boy in the fable about the emperor's new clothes and say that I see no subject. There are technical problems in abundance, intriguing observations, puzzling phenomena, and some scraps of theory about specific features, but not yet the kind of secure foundations and body of theory which turn a collection of particular problems into a subject". Despite these rather scaring prospects, Batchelor succeeded in contributing to the field, as did several researchers in the Netherlands.

## • 5.3.1 EARLY DEVELOPMENTS

For the *Second Report on Viscosity and Plasticity* of the Viscosity Committee (1938; see § 5.2), Burgers had written a contribution on 'the motion of small particles of elongated form, suspended in a viscous liquid'. In 1940 he took up the subject of suspensions again when he had noticed that others had applied some of his formulae but found discrepancies between theoretical and experimental results. Burgers discussed this problem in his paper and suggested new formulae.

Apparently, Burgers kept thinking on the problems related to suspensions despite the toughness of the topic. In 1941, he contemplate on diffusion and in a paper published in 1942, he again considered suspensions. It was written, as he remarked to one of his correspondents, "with great enthusiasm and with the hope of straightening out some questions". However, "with its continuation more and more problems appeared which baffled me". Therefore, he had to admit that he was not very satisfied with the results. The problems encountered by Burgers essentially concerned the lack of absolute convergence of the sum of the separate effects of an indefinitely large number of falling spheres on a given sphere.

Though Burgers felt somewhat disappointed, today it is recognized that he was one of the first who found a method to calculate an 'effective viscosity' of suspensions. For his result he had used the theory of Oseen which he had studied so deeply in the 1920s (see § 5.1.3). From the references in his papers, it becomes clear that Burgers was well acquainted with the literature related to colloids. He also knew Kruyt in Utrecht well (see § 4.1.4) but there are no traces of any collaboration with Kruyt or one of his co-workers (after the Second World War, Burgers' brother Willy would start to give lectures on colloid theory in Delft). It seems that during the war an experiment on sedimentation in 'artificial turbulence' was started in Burgers' laboratory but due to the circumstances, was never finished. As for the theoretical approach of dispersions of particles

# TEIJIN ARAMID AND THE TWARON ARAMID FIBER
## HANS MEERMAN, TEIJIN ARAMID

The first artificial fibers that could match wool and silk were produced at the end of the 19th century. In the course of the 20th century the focus was on the production of technically advantageous - meaning strong and rigid - fibers. Experience made clear that the molecular orientation of a polymer with sufficient molecule length is important to the strength and rigidity of fibers. In about 1970 it was understood that polymers containing sufficiently rigid and stretched polymer chains will align at higher concentrations when dissolved (liquid crystalline or LCP behavior) causing the solution viscosity to remain within reasonable limits. During the research into polymers with LCP behavior, poly-paraphenylene terephtalamide (PpPTA) turned out to be a good choice for further development for practical and economic reasons. The chain alignment taking place during the dissolved polymer flow leads to a high degree of crystallization in the resulting fiber.

The high degree of crystallization produces the fiber that is commonly known as Twaron, its high strength and rigidity being of importance to many high performance technical applications. As shown in the illustration, the formation of a high degree of alignment of dissolved polymer chains has been visualized by a dedicated visualization technique (crossed polarizers). The flow is recorded just before entering the extrusion capillary in which the fiber is created. The definitive alignment is realized by further stretching the fiber underneath the extrusion capillary and by fixing the alignment and fiber shape in water. The fixing process for shape and alignment happens within a very short time and constitutes the core of the spinning process used for the large-scale production of the super fiber under its brand name Twaron in Emmen.

Teijin Aramid is a subsidiary of the Teijin Group and a global leader in aramid production. You will find the aramid fiber Twaron, Teijinconex, Technora or the ultra-strong (UHMW-PE) Endumax wherever strength, safety, heat resistance or light weight is needed. Teijin's products can be found globally in many diverse applications and markets, such as the automotive, ballistic protection, marine and civil engineering, protective clothing, rope and optical fiber cable manufacturing, and the oil & gas industry.

We use our high-performance materials to offer a wide range of products. Our specialist knowledge and many years of experience enable us to work continuously on new products and solutions. We are continually busy creating added value for our customers, strengthening their competitiveness with our innovations. Therefore, we invest more than 4% of our turnover in research and development at our dedicated establishments in Arnhem (the Netherlands), Matsuyama (Japan), Iwakuni (Japan), Wuppertal (Germany) and Shanghai (China) Thanks to our regional R&D centers, we are always 'close' to our source and able to be of service to our customers on a local level. It is actually our conviction that our best solutions are developed in the cooperation with our customers. By co-creation we arrive at solutions that help our customers excel in their markets. Our motto is: think global, act local. We observe and anticipate on global changes and developments, ensuring that we are and will remain to be the world's number one player in aramid manufacturing. With our world-wide sales and marketing organization, we are always close to our customers and are able to speak the language of our customers.

The aramid production and chemical processes take place in the Netherlands (Delfzijl, Emmen, and Arnhem), in Japan, and in Thailand. The world's biggest aramid plant is located in Emmen, the Netherlands. Teijin Aramid, a global company of 1,700 employees, established its headquarters in Arnhem.

  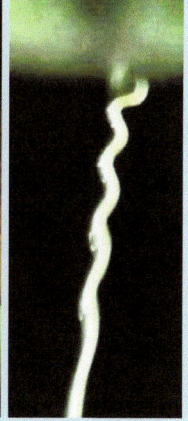

↑Flow of a liquid crystalline aramid solution right before entering, respectively, one and three extrusion capillaries, and an unstable flow exiting one of the extrusion capillaries. The alignment of the PpPTA chains is visualized by two crossed polarizers. The capillary diameter is 100 μm. The distance between the extrusion capillaries is 0,5 mm. (from: Drost (2015), PhD research supported by Teijin Aramid; courtesy of Teijin Aramid)

→A facility in Kramers' laboratory around 1950, for the 'merging of gas and liquid'. (from: Reynhart (1951) / courtesy of Shell)

↑Since its early days researchers at the Kramers Laboratory have been studying fluidized beds. Here professor Rob Mudde is standing next to a so-called 2D dry matter fluidized bed in 2009. (courtesy of TU Eindhoven archives / photo by Bart van Overbeeke)

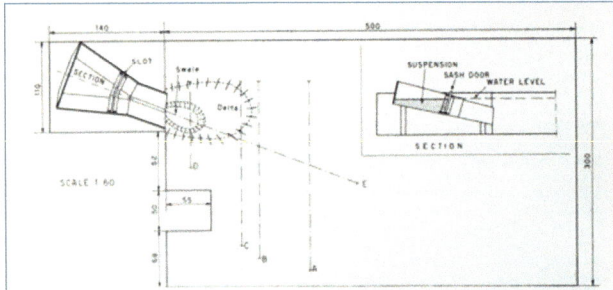

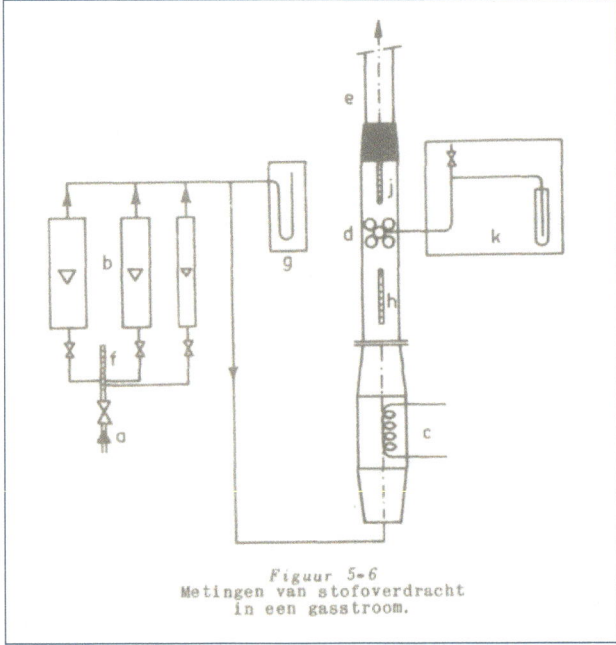

*Figuur 5-6
Metingen van stofoverdracht
in een gasstroom.*

↑Kuenen started his career at the Geological Institute of the University of Leiden and his first experiments on turbidity currents were probably performed in this basin (about 6 x 4 m) which had a wave generator. Kuenen was much inspired by Berend Escher, a pioneer in geological experiments. Escher did, e.g., experiments with standing waves and studied their influence on an artificial beach. In Groningen Kuenen used a somewhat smaller basin in which a suspension was released on a flat slope. (from: Leidsche Geologische Mededeelingen, IX, 1937; from: Hough (1951) / courtesy of Soc. for Sedimentary Geology)

↑Kramers' second PhD student was Dirk Thoenes, who later became professor in Twente and Eindhoven and an expert in chemical reactors. His thesis (in Dutch), published in 1957, was entitled *Mass transfer by means of flow through a solid bed of granular material*. Thoenes did experiments with gas flows which were sent through a vertical tube and took with it an evaporated liquid from a porous sphere in an elementary model of a granular bed (near d). (from: Thoenes (1957))

in liquids, Burgers got assistance from a PhD student from China. In 1947 Tchen Chan-Mou defended his thesis entitled Mean value and correlation problems connected with the motion of small particles suspended in turbulent fluids.

To Burgers' successor Hinze, two-phase flows were already well-known when he started in Delft in 1956. He had published a formula for droplets (see § 5.1.5) and would continue to write on dispersions and particles in turbulent flows during the 1950s and 1960s. In 1959 the Laboratory for Aerodynamics and Hydrodynamics would see its first experimental facility for two-phase flows: a closed water circuit into which granular material could be dosed for studies on the flow of dispersions.

In the meantime another laboratory in Delft had built several facilities for research on two-phase flows. Since these flows are very common in many industrial processes, as in the oil industry, it was only natural that Kramers in his Laboratory of Physical Technology would start experiments on process machinery such as chemical reactors and fluidized beds. Kramers and his co-worker Westerterp published the first monograph on reactors ever, Elements of chemical reactor design and operation, in 1963. From the 1960s bubble columns also became familiar facilities in the Kramers Laboratory.

## • 5.3.2 DISPERSIONS

Among the several two-phase flows which can be studied, those in which particles, droplets or bubbles are dispersed have been the most studied by academic scientists in the Netherlands. One of the aspects of these flows which makes them hard nuts to crack is the 'hydrodynamical interaction' between the dispersed elements (which has been called Stokesian dynamics). For droplets and bubbles the processes are even more complicated since these can 'merge' (the coalescence of droplets has been one of the topics of the groups of Lekkerkerker in Utrecht and Bonn of the University of Amsterdam).

### DISPERSIONS OF SOLID PARTICLES

One of the topics of two-phase flows on which Batchelor, whom we have met in the introduction, has published is the rate of sedimentation of small solid particles. Beenakker en Mazur of the University of Leiden (see § 4.1.4), however, criticized his results and showed that this rate depended on the shape of the vessel in which the sedimentation took place, something which Burgers had already mentioned (but could not prove) in 1941.

Dispersions of particles (and bubbles) in liquids also attracted and challenged scientists normally working in other fields of physics. For example, Ubbo Felderhof (1934) was known for his work in plasma physics and statistical physics but in

1991 published a paper entitled 'Virtual mass and drag in two-phase flow' in the Journal of Fluid Mechanics. (Some years later he was working on the theory of swimming.)

Sediments had long been a field of interest for hydraulic engineers (see e.g. § 4.1.1). One of the most intriguing aspects was the formation of ripples on sand bottoms. In 1960s experimental research related to this was done in Hinze's laboratory. In one of the wind tunnels mean velocity and wall shear stress were measured on artificial ripples with pressure holes.

Several decades later, sand transport by water over ripples in shallow seas was investigated in the group of Hulscher (originally a theoretical physicist) at the University of Twente. This transport has been one of the largest knowledge gaps in the modelling of sand transport in coastal seas. From full-scale experiments and modelling it appeared that the ripples strongly influence the boundary layer structure and turbulence intensity near the bed and have a great influence on the sand transport. In the 1990s Hulscher had already found an explanation for the emergence of sand ripples and sandbanks. Ripples were also studied by Martin van Hecke at the University of Leiden (the Lorentz Institute) and Huib de Swart at the University of Utrecht

Somewhat related to sediment phenomena are the so-called turbidity currents, akin to avalanches of sand and water on the bottom of seas and oceans. A pioneer in the experimental study of these currents was professor Philip Kuenen (1902-1976), marine geologist at the University of Groningen. While still a young scientist in Leiden, he did experiments with flumes and wind tunnels on the rounding of stones and sand grains. In 1937 he started experiments to determine the origin of so-called submarine canyons. In his Geological Institute in Groningen he built a sloping flume on which a large amount of sediment was suddenly released. In the 1950s these experiments became well-known internationally and in 1960 Hinze in Delft tried to give a theoretical foundation of these currents. More than half a century later experiments on this phenomenon were undertaken by scientists from the Department of Geosciences in Utrecht. In the Eurotank of TNO they were able to mimic the formation of underwater gullies.

Transport, or displacement, of water and dispersed material like sand or clay can be found in many situations: the breaching of dikes, unstable breaching of sand during dredging, transport of slurries through pipelines, etc. Dispersed solid particles also occur in the mining industry. There, liquids are used to separate granular materials of different densities. The pulsating liquid flow which is used for this, is called a jig. Researcher at the Department of Mining Engineering in Delft have been successful in developing a jig with which every periodic wave desired can be generated

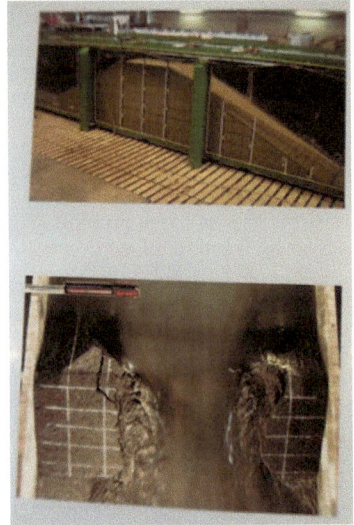

←Clay and water can be a dangerous combination. In 2006 Yonghui Zhu, PhD student from China, published his thesis after his research done in the Waterlab in Delft to study the breach growth erosion process in clay-dikes and to obtain data to test a mathematical model. Despite the fact that dike breaches have occurred in the Netherlands (1953), in China and many other countries for centuries, the knowledge about the processes involved was still poor and the models were not very advanced. In general, the behaviour of granular materials, of which clay is just one example, is still poorly understood (see § 5.2). (from: Zhu (2006) / courtesy of dr Yonghui Zhu)

↑The behaviour of sediment is one of the main topics of the research at the Dredging group in Delft (see also § 6.2.6). One of their experiments, related to offshore deep-sea mining, is shown here: on a flat slope sand-like particles are dispersed. (courtesy of Delft University of Technology / photo by Ivar Pel)

→Today bubbles are also applied outside the chemical world. Deltares and RWS have developed a 'bubble screen': bubbles arise from the bottom, in this case in the Krammen Locks in Zeeland. This minimizes the penetration of salt water from the sea into the rivers, without hindering ships. (courtesy of Rijkswaterstaat / beeldbank.rws.nl / photo by Kees Jan Meeuse)

and this has led to a strong increase in the efficiency during separation.

Numerical simulation of these processes has been quite a challenge but the growing amount of experimental data helps to improve the models. One of the many attempts in modelling two-phase flows is the 'discontinuous Galerkin finite element method' developed in the 2000s at the Department of Applied Mathematics in Twente for shallow two-phase flows.

### DISPERSIONS OF GAS BUBBLES IN LIQUID

As was said before, dispersions of bubbles have been in the (academic) laboratories for decades in the form of bubble columns. When Van Wijngaarden came to the University of Twente in 1966, he started to study bubbles, the beginning of a long tradition at this university. He started to do theoretical work on bubbles whose behaviour had become known to

him when studying them in the wake of ships at the NSP in Wageningen where he had worked from 1962. His paper 'On the equations of motion of liquid and gas bubbles' of 1968 became well known. He was also a pioneer in modelling the interaction of bubbles and he did important work on the noise generated by cavitation. Van Wijngaarden managed to formulate a nonequilibrium mathematical model for the description of wave processes in liquids with gas bubbles. This model was later named after him and two Soviet scientists: the Iordansky-Kogarko-Van Wijngaarden Model.

Besides the behaviour of swarms of bubbles in liquids, researchers in Twente have also studied single bubbles. In a PhD thesis published in 2007 the so-called Leonardo's Paradox was demystified: why do bubbles stop to rise in a straight line from a certain diameter and start to zigzag? In Delft bubbles have also been investigated for decades.

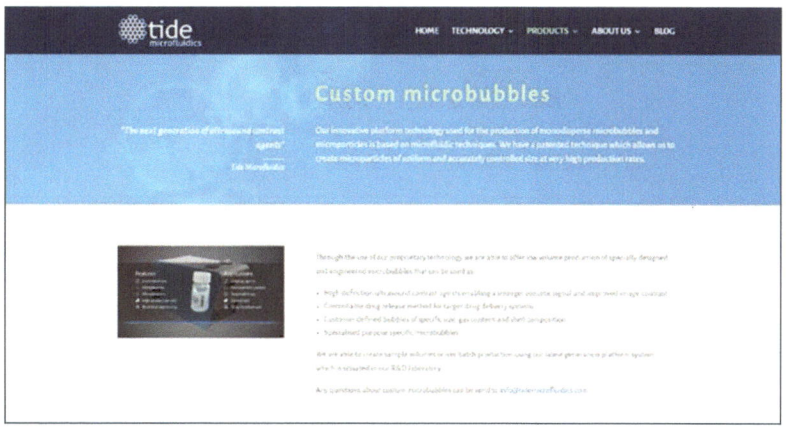

↑Bubbles have become business. The Tide Microfluidics company offers tiny particles and bubbles for medical purposes. Its origins can be found in the research groups of Van Wijngaarden and Lohse. (courtesy of Tide Microfluidics BV / Wim van Hoeve)

↑Colloidal suspensions have traditionally been studied in physical chemistry but in the 1990s physicists in Amsterdam also started to study them. Observing the scattered light of a laser, they discovered a new phenomenon. In these suspensions all particles are moving around. At the front of each particle a kind of bow wave is present and the liquid there is pushed away towards the wake of the particle. The velocity fields around all the particles influence each other. In this experiment it became clear for the first time that the velocity around one particle is shielded by the fields of all particles around it. The 'cage' of particles in which the central particle is located, determines the diffusion of this central particle. With this phenomenon they could explain why the rate of sedimentation observed in experiments is always lower than the theoretical rate. (courtesy of University of Amsterdam / Daniel Bonn)

←The 'turbulence box' in Eindhoven used to have eight speakers but a later version (shown here) has twenty. Four cameras are used to track the droplet motion. (courtesy of Eindhoven University of Technology / Rudie Kunnen)

In the Laboratory for Physical Technology (the Kramers Laboratory) this led to the construction of a unique bubble column (see § 6.2.6). Fundamental research on single bubbles and the coalescence of two (and more) bubbles has been done by Chesters in the Laboratory of Aero- and Hydrodynamics.

Connected bubbles can form foams. In this context one very specific PhD thesis has to be mentioned: in 1989 Lex Ronteltap, working at Heineken Breweries, defended his thesis entitled Beer foam physics. Ronteltap had studied four phenomena: bubble formation, drainage, disproportionation, and coalescence.

### DISPERSIONS OF LIQUID DROPLETS IN GAS

Examples of dispersions of droplets can be found in sprays. The breakup and droplet dispersion of sprays is an import-

ant aspect of many processes. One of them is the combustion process, where the behaviour of the droplets influences the efficiency and pollution of engines. Researchers of the Turbulence and Vortex Dynamics group in Eindhoven have used glowing sprays to find out what happens to droplets in a turbulent flow. To measure the effect of turbulence, they built a 'turbulence box' in which turbulence is created by air jets pumped by large speakers. These speakers are driven by (colored) noise signals.

### • 5.3.3  TWO-PHASE FLOW AND THE OIL & GAS INDUSTRY

A large part of the research on two-phase flows in the Netherlands has been, and still is, somehow related to processes which take place in the oil and gas industry. Shell especially

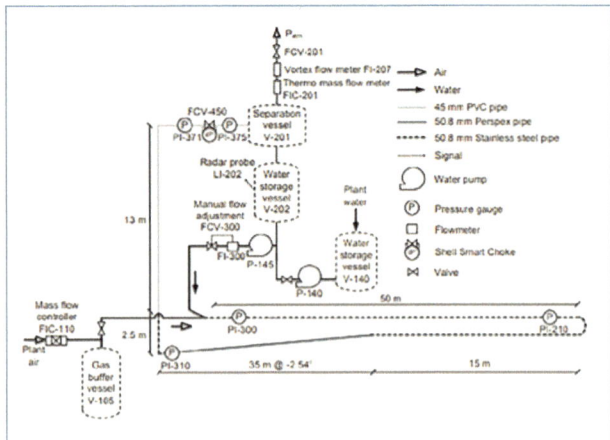

↑ 'Severe slugging' research as presented in a PhD thesis published in 2012 (supervised by professor Rob Mudde of the Kramers Laboratory) including experiments in this large facility (the Severe Slugging Loop) at the Shell Technology Centre in Amsterdam in which slug flow in a flowline-riser system could be investigated (from: Malekzadeh (2012); courtesy of TU Delft).

Shell has stimulated research in this field, not only in their own laboratories in Amsterdam and Rijswijk, but also at the universities.

During the 1960s, 1970s and 1980s fundamental research on two-phase flow phenomena seems to have been given almost free hand at Shell's KSLA. To give an example: around 1970 the development of bubbles in a boiling liquid above a glass surface was filmed and measured. This gave the researchers important information about the rate of heat transfer in processes where boiling was present. After a reorganisation of the KSLA at the beginning of the 1980s (leading to a reduction of research space and staff) the number of joint projects between Shell and the academic world started to grow.

Since the beginning of the 70s, much two-phase (or multi-phase) flow research was aimed at understanding the transport of oil, gas, water, and solids though pipelines. This required much laboratory experimentation, field data analysis, and model development. The problems were and are challenging as many physical phenomena (like turbulence, interface instabilities, rheology, thermodynamics) are involved.

## EXAMPLES OF THE RESEARCH FOCUS IN HEAVY-OIL TRANSPORT

A large part of the remaining oil in underground reservoirs is very viscous and this property causes problems during its recovery, transport, and processing. Therefore, methods need to be developed to reduce the pressure drop encountered during the transport of such oil type. One well-known method to make this possible is the addition of some water to the viscous oil. Under certain conditions

(such as a sufficiently high flow rate), this gives the so-called core-annular flow regime in the pipeline. This means that the oil flows in the core of the pipe, which is surrounded by a thin layer of water in an annulus along the full perimeter of the pipe. Now the water works as a 'lubricant': the frictional pressure drop along the pipeline is thus determined by the low water viscosity, and not by the high oil viscosity. One of the essential phenomena which takes place during this type of flow and which has been subject of a lot of research are the waves that are formed at the interface between oil and water. Around 1990 this kind of flow was studied at KSLA in various facilities, such as a tube of 16 m length and 5 cm diameter and a tube of 1000 m length and 20 cm diameter. Ultrasonic sensors were used to measure the thickness of the water layer.

Recently one of the facilities used in the past at the Shell laboratory has been rebuilt at the TU Delft, and the core-annular flow research has been continued. In addition to its application, core-annular flow is a very interesting configuration for fundamental research on two-phase interface behaviour, particularly the interaction between the turbulence and waves. New compared to the previous research is that now much larger computational resources and better numerical methods are available. But the application of simulation methods to interface flow is still a challenge. Besides the core-annular flow (which is a liquid-liquid two-phase flow), the so-called slug flow regime in gas-liquid transport in pipelines has been the topic of research at various Dutch laboratories. In the Laboratory of Aerodynamics and Hydrodynamics and in the Kramers Laboratory, both in Delft, research was carried out on slug flow in pipes at various inclinations, including the extremes of a vertical and

→ Separation is an important process in the oil industry. This picture shows the laboratory of the Romico company, which was founded by professor Bert Brouwers. Brouwers started his career at the Ultra Centrifuge Laboratory of UCN/Urenco in 1972, then he worked for Shell, and in 1986 he became professor of Thermal Engineering in Twente and in 1998 he became professor of Process Technology in Eindhoven. During his time at the University of Twente he devised the Rotational Particle Separator (RPS) which comprises patented methods for separating micron-size particulate matter from fluids. While still working at the university, he started his own company to develop practical versions of the RPS and to sell these. (courtesy of ROMICO Hold avv / Bert Brouwers)

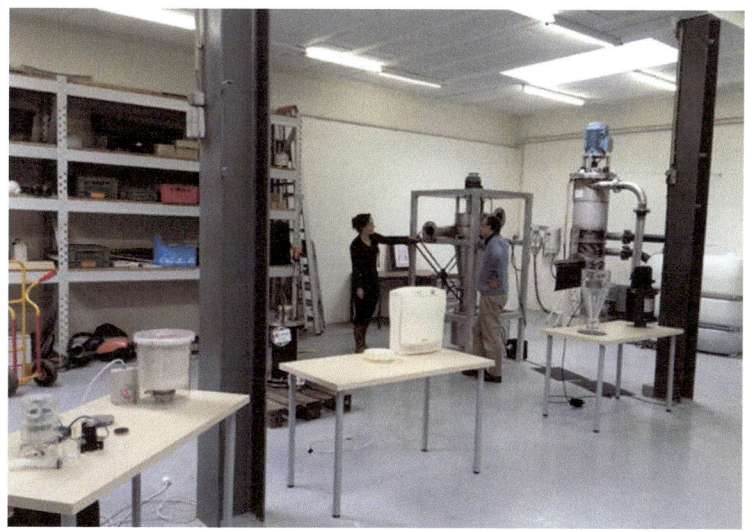

← PhD student Dries van Nimwegen measureed the formation of foam to prevent slugs in vertical two-phase flow at the Kramers Laboratory in Delft around 2015. Slug flow research in this lab dates back to the 1990s when a vertical pipe of 17 m length was installed. (courtesy of NAM / photo by Levien Willemse)

a horizontal inclination. Different slug types were measured, such as Taylor bubbles, hydrodynamic slugs, and growing slugs (see also § 6.2.6). Such research is still ongoing, but now even under more complex conditions, such as the measurements for the effect of a surfactant (foam) to prevent hydrodynamic slug formation in vertical pipes. Slug flows have also been measured at one of the TNO laboratories, for example the fluid-structure interaction due to bend forces generated by hydrodynamic slugs.

Through the years much slug flow research has also been carried out by Shell researchers at their laboratories in Rijswijk and in Amsterdam. In Rijswijk Shell used the Donauloop for three-phase flow measurements. In Amsterdam researchers could use the Severe-Slugging Loop (SSL) in which severe slugging has been measured, which can occur in a flow-line riser system (which is a horizonal or slightly inclined pipeline followed by a vertical pipe). During a severe slugging cycle, the full height of the riser is filled with liquid, as well as part of the flowline. Once the liquid reaches the top of the riser, liquid leaves the top at high velocity, which is a so-called liquid surge, followed by a gas surge. The experiments were used to derive simple correlations for the onset of a severe slugging cycle, as well as for developing dedicated control methods, in which an actuated valve at the top of the riser is used to prevent the severe slugging cycle or to mitigate the liquid and gas surges.

More recently there has also been much interest in the numerical simulation of slug flow. This means solving the time-dependent, one-dimensional (in space) conservation equations for the gas and liquid dynamics, requiring advanced numerical methods for so-called slug-capturing. This work is carried out at in the Applied Mathematics department of the TU Delft, the Centrum voor Wiskunde en Informatica (CWI) and at Shell.

Much of the core-annular flow and slug flow research as described above was or is carried out under the supervision of the professors Ooms, Oliemans, Mudde, Vuik, and Henkes.

During oil recovery from reservoirs a mixture of oil and water is brought to the Earth's surface. There the oil is separated from the water. The larger the oil droplets the easier the separation will be. In 2000 a PhD thesis on the effect of restrictions in pipelines (which are due to the presence of various valves) on the break-up of droplets was published. The experiments for this research had been done in the Dietz Laboratory of the Department of Applied Earth Sciences in Delft.

Foam also has its place in the world of oil recovery. Oil from oil-gas reservoirs is initially produced by the natural driving mechanism of the pressurized reservoir, which gives the primary production. During the secondary recovery phase water or gas are injected to maintain the pressure in the reservoir. After these two phases, more enhanced recovery techniques are used. One of these is the use of foam, which is an idea that dates from the 1950s. Foam has several applications in oil production, e.g., in drilling where foam can transport cuttings to the surface, and in deliquification of gas wells. In 2012 a PhD thesis on the modelling of foam for oil recovery was published under the supervision of professor William Rossen of the Department of Geoscience and Engineering in Delft.

# FLUID FLOW RESEARCH IN SHELL THROUGH THE YEARS
## RUUD HENKES & PETER VEENSTRA, SHELL PROJECTS & TECHNOLOGY

Hinze carried out pioneering work for Shell during his time in the Proefstation Delft in the years 1935 to 1956. In the 1950s his work was continued by J.G. Van de Vusse at KSLA who worked on the mixing scaling rules in process equipment. His scaling theory was tested at full scale, as is clear from the example of an educator mixer that was placed in a tank of 12,000 m3. The mixing insights of Van de Vusse are still valued in the mixing engineering community. In later years research in Shell has paid much attention to mixing, due to its importance in equipment or processes like side entry mixers, bubble columns, jet mixers, ejector mixers, in-line mixers, distillation columns, trickle bed reactors, drilling, well, pipelines, separators, etc.

In the 1960s and 1970s J.A. Wesselingh studied, among others, the scaling of storage tanks. An example is a small-scale experiment with a tank model in which the mixing of black and white miscible fluids is followed over time. This work led to various engineering design rules for fluid mechanics (such as mixing times for miscible fluids and de-mixing or stratification times for immiscible fluids), solid mechanics (such as selection of seals, bearings, and shaft), and process control. The design rules are still in use for many of the processes and processing units in refineries and chemical plants. Hans Wesselingh left Shell in 1976 to become lector and later professor in Separation Processes at Delft University; he moved to the University of Groningen in 1989 to become professor in Thermodynamics and Separation Processes.

Over the years a number of university professors in fluid flow or process engineering have spent parts of their career in Shell. In addition to professor Hinze and professor Wesselingh, as already mentioned, other well-known examples include the professors Charles Hoogendoorn, Gijs Ooms, Harry van den Akker, René Oliemans, and Dirk Roekaerts. During their time in Shell they contributed to a variety of research topics in applied fluid flow. The results of their work have been described in many Shell internal reports, but some results were also presented and published at conferences, and in international journals.

Up to the 1960s fluid flow research in Shell was mainly focussed on the "downstream" part of the oil and gas business, which are the refineries and chemical plants. Thereafter there was also much interest in the "upstream" part, being the exploration and production of oil and gas, which includes the flow from the reservoir, through wells, pipelines, and risers, to production platforms or gas plants. Over the past few decades much research was done to understand the multiphase flow in pipeline systems. Starting with the oil crisis in the beginning of the 1970s, it became attractive to transport the hydrocarbons in a single multiphase flow pipeline, instead of using pipelines for either single phase gas or single-phase oil transport.

At the KSLA lab pipeline experiments were carried out to generate flow-pattern maps for two-phase flow transport in pipes. Among the lab facilities, there was a 130 m long, 50 mm diameter water/air loop, that was moved from Shell in Amsterdam to the Kramers lab at Delft University in 1999. Another valuable source of data also was found in Shell's pipeline facility at Bacton in the UK, which was directly fed by gas and condensate from the field. Measurement equipment included pressure sensors, flow visualisation through a looking glass in the pipe, liquid accumulation (liquid holdup) measurements through a gamma density meter, liquid droplet entrainment through isokinetic sampling. Mechanistic models were derived to describe the important multiphase flow parameters: flow regime, pressure drop, liquid holdup. And the lab and field data were used for the model validation. This has led to a model called the "KSLA method" for multiphase flow in pipelines. The KSLA lab was later renamed into SRTCA (Shell Research and Technology Centre Amsterdam) and is now called STCA (Shell Technology Centre Amsterdam). The flow models were also renamed into SRTCA method and are now simply referred to as the "Shell Flow Correlations".

The Shell Flow Correlations are presently still used for the multiphase flow design and improved operation of gas, condensate, water, and solid particles in offshore and onshore pipelines. The correlations are embedded in larger steady state and dynamic simulation packages. A new development is that the dynamic pipeline models are brought on-line. This means that the model serves as a Digital Twin of the actual system that is in operation. The model is fed by field measurements for quantities like the flow rates, inlet temperature, and arrival pressure, which serve as boundary conditions. The model can provide data for quantities that cannot be easily measured, such as the liquid holdup in the pipeline. In this way the model can be used for liquids management. It can tell when and how large pockets of liquid (so-called slugs) propagate through the pipeline, and whether such slugs still fit the capacity of the downstream separator or slug catcher. The model can provide direct monitoring but can also forecast for the next 24 hours or so. Since the 1990s, Computational Fluid Dynamics (CFD) has grown enormously as a research and design tool for fluid flow problems. In the Shell lab researchers have been and are still working on the numerical algorithms and on the physical closure correlations. This has led to Shell in-house tools based on e.g., the Lattice-Boltzmann Method (Cellular Automates), Large-Eddy Simulation (LES), and Smoothed-Particle Hydrodynamics (SPH). Increased use is also made of third-party CFD tools, like Fluent and STAR-CCM+. As partner in various Joint Industry Projects (JIPs), Shell contributes to enhancing the functionality of CFD packages. For solving practical problems, often both momentum, mass and heat transfer are important at the same time, as well as the kinetics of chemical reactions. This asks

↑Full scale mixing experiment as carried out by Van de Vusse in 1953. (courtesy of Shell)

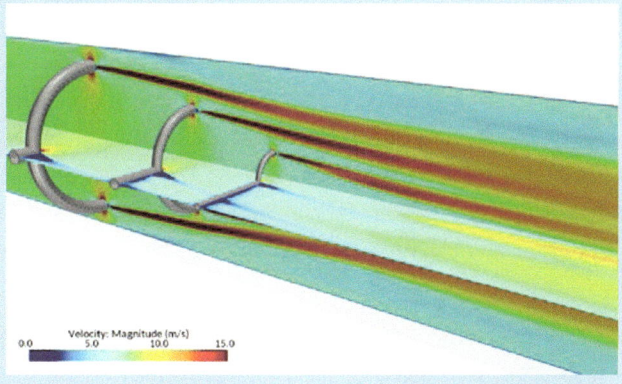

↑CFD simulation (carried out by Shell in 2017) to determine the optimum injection of oxygen in a fabrication process of chemicals. (courtesy of Shell)

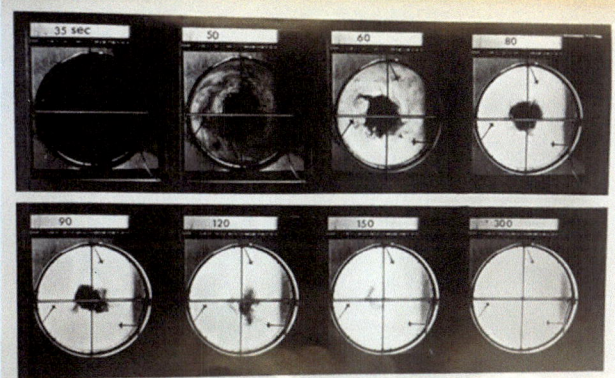

↑Results of a mixing experiment as carried out in the Shell lab by Wesselingh in 1967. (courtesy of Shell)

↑Shell Technology Centre in Amsterdam as it is today. (courtesy of Shell)

much of the computational tools, both from a numerical (computational speed and robustness) and physical view point. CFD is used for a wide range of problems, such as efficiency improvement in chemical reactors (e.g. the effect of internals), erosion of wall materials due to the impingement of solid particles in the fluid flow, fluid-structure interaction (e.g. to find the root cause of vibrations of equipment), and mixing in storage tanks.

For Shell fluid flow research will also remain important in the coming decades. This will be a combination of in-house research and leveraging through collaboration with national and international research institutes. Developments in Information Technology will also provide great opportunities to make better use of the data available from field operations. Experience from the past years in fluid flow modelling can be used to realise fast benefits from further Digitalization and Machine Learning. Having a fluid flow simulation model with direct coupling to operational data can provide better

and safer operational performance, once algorithms with artificial intelligence have developed to a state that they can provide early warnings of anomalies in the monitored processes or units.

Within the energy industry, fluid flow is seen as an enabling technology. It is clear that it has been of importance for the maturation of the oil and gas industry. The heritage is a wealth of engineering design rules and dedicated design tools. Fluid flow will remain an important enabler when designing the new energy future and new technologies for chemical products. Possible research topics are plenty, for example: wind turbines, solar boilers, hydrogen production and transport, electro-synthesis of chemical products. More than in the past, fluid flow needs to become fully integrated into other technical areas (e.g., process engineering, materials, structural mechanics, information technology) to make an impact in the coming years.

95

© Springer Nature Switzerland AG 2019
F. Alkemade, *A Century of Fluid Mechanics in The Netherlands*, https://doi.org/10.1007/978-3-030-03586-0_6

# 06

## RESEARCH IN FLUID MECHANICS: APPROACHES

For centuries fluid mechanics was mainly approached in two different manners: by using mathematics (and intuition) or by doing experiments (in or outside of a laboratory). Only from the 1950s it became possible to take the third way of approach: the numerical one; first mainly by hand, later using analogue and digital computers. In this chapter Dutch examples from all three types of approach are presented.

# 6.1 THEORETICAL AND MATHEMATICAL APPROACHES

As in many other fields of physics and engineering, the theoretical approach is often a tough one and the number of real breakthrough results is small. The boundary layer theory and the lifting-line theory for 'real' wings developed by Prandtl may be called such results, but there were not many others like them in the period between 1904 and the outbreak of the Second World War. The same can be said of the post-war period.

The method most often used to approach flow problems theoretically is, of course one could say, to use tools from mathematics. All flows are described by the Navier-Stokes equations (usually indicated by NS) and equations can be 'attacked' by means of methods which have been developed by (pure) mathematicians. The big issue which overshadows all these attempts is the fact that the NS are nonlinear. This nonlinearity bothered the physicists and hydraulicians of the 19th century, it bothered Lorentz when he tried to do calculations on the behaviour of the Zuiderzee in the 1920s (see § 3.4.1), and it still bothers all those who try to understand and predict flows by just using pen, paper, and their mathematical toolbox.

Lorentz chose an attractive (but to some doubtful) 'simplification' of the nonlinear equations: he linearized a component in the equations related to friction, and could prove that this did not hinder him in obtaining useful results. Some years later Dronkers, a mathematician working at Rijkswaterstaat, came up with a more accurate mathematical model of the friction term (see § 6.3.1).

Somewhat surprisingly perhaps, many years later this Lorentz linearization was studied again by another Dutchman, Sjef Zimmerman who worked at the NIOZ (see § 4.2) in the 1980s and 1990s. Zimmerman worked on so-called renormalization techniques in fluid mechanics. With these techniques he was able to find approximate solutions of non-linear partial differential equations by following a 'recipe' which he had abstracted from applications of adjoining fields of physics. With his method Zimmerman could show that Lorentz's approach can be justified.

### BURGERS AND THE THEORETICAL APPROACH

When Jan Burgers started in Delft in 1918, he had no experimental facilities at his disposal and thus he was simply obliged to take the theoretical approach. For this he was well prepared, as we have seen in § 3.1: among his tutors were Lorentz himself as well as Ehrenfest. One of the mathematical techniques he applied was conformal mapping, and he became an expert in this. In the *Annual Review of Fluid Mechanics* of 1975, Burgers would recall: "Conformal mapping, as a means for obtaining contours of airfoils and of propeller blades, for a long time was a dominant subject for me. I was particularly interested in looking for the simplest formulations to be used in my lecture course. But it was also evident that everywhere in fluid dynamics attention should be given to vortex motion. Far more knowledge was required than was given in H. Lamb's Hydrodynamics, notwithstanding the importance of Lamb's work. [...] I began to see that transport of vortex motion, partly by convection through the general flow field and partly by diffusion as a result of viscosity, was of decisive importance in many cases, and I formulated a relation between the resistance experienced by a body and the momentum or impulse of the vortex system generated." In fact, Burgers' very first paper on fluid mechanics, in 1920, is entitled 'On the resistance of fluids and vortex motion'.

Both conformal mapping and his knowledge of vortex motion would later appear useful in several investigations: the former was applied to the calculation of flows along fans in pumps (see § 6.3), while vorticity appeared to play an important role in the calculation of flow resistance and in the application of Oseen's theory.

Carl Oseen was a Swedish theoretical physicist who had derived equations, later named after him, which describe the flow of a viscous and incompressible fluid at small Reynolds numbers, somewhat similar to a Stokes flow. Burgers, amongst others, was intrigued by Oseen's results and he started a correspondence with Oseen which lasted from 1919 up to 1946. He even gave lectures on Oseen's theory for his students. As Burgers recalled in 1975: "In subsequent years I continued to work upon Oseen's approximation and its relation to Prandtl's theory. It appeared that Oseen's equations for the flow called forth by exterior forces acting on a fluid could be used for the description of the vortex system produced by a lifting system, and so gave a direct connection with Prandtl's theory of the finite wing. Later, I used Oseen's equations for the calculation of the resistance experienced by small particles in slow motion, at Reynolds numbers far below unity."

When he started to work on turbulence around 1930, Burgers also tried the theoretical approach. Like several of his foreign colleagues he started to study statistical theories of turbulence, but he soon discovered several difficulties and felt dissatisfied with his own papers on this matter. To the English physicist George Trubridge, who in the 1930s wanted to write a thesis on 'Burgers' theory of turbulence', he wrote in 1933: "[These difficulties are] intrinsically connected with the statistical method used in these papers, and it is not peculiar to the hydrodynamical problem. I happened to find certain systems of simultaneous differential equations which show properties analogous to those of the equations of fluid motion in so far as regards the existence of solutions representing a state of 'turbulence', although they are much simpler."

Thus, in the 1930s Burgers took yet another road to treat

turbulence. He decided to restrict his attention to model problems, with which the essential aspects of turbulence could be treated. As he explained in his valedictory lecture of 1955, he became convinced "that dissipative systems are essentially different from conservative systems. I thought that it would be necessary therefore to study the behaviour of dissipative systems, and that since the Navier-Stokes equations are so refractory, it might be helpful to replace them by a more elementary equation. It was then that I took as example the equation

$$\partial v / \partial t = U v + v \, \partial^2 v / \partial y^2 - 2 v \, \partial v / \partial y$$

and I prepared an extensive investigation of this and a few similar equations in a paper published in 1939."

The first sketches for this paper date from 1936. A year later he showed that it was possible "to illustrate the conception of correlation and the equations describing the decay of free turbulence" with his new model. To Burgers, a new road seemed to have been opened which would finally bring progress. In his valedictory lecture, he remembered: "Especially, I started to free myself from the spatial character of the flow; thus, I was not troubled by the complicated geometrical properties of the vortices; also I abandoned the continuity equation."

After the War, Burgers got acquainted with developments in the U.S.A. and Britain. He realized that his treatment of turbulence had taken a direction which differed strongly from those taken by others like Kolmogorov and Heisenberg. Nevertheless, he still felt convinced of the value of his work. Finally, the equation which Burgers would study most in his work on turbulence, became known as the Burgers equation:

$$\partial u / \partial t + u \, \partial u / \partial x = v \, \partial^2 u / \partial x^2.$$

To Burgers, they contained the essential features of turbulence: dissipation and nonlinear inertia. In a paper of 1954 he considered wave-like solutions of his equation which, he thought, could serve to illustrate the interaction of shock waves, which he had been working on in the 1940s. For this purpose, he applied a substitution which is known as the Hopf-Cole transformation and which turns the nonlinear Burgers equation into the linear heat equation (Burgers had learnt about the transformation from J.D. Cole himself when he visited the U.S.A. in 1949).

The attractive properties of the Burgers equation are the facts that it is analytically treatable and that it contains essential ingredients of turbulent flow. As Burgers remarked himself in 1954, it is "the simplest analogue of the hydrodynamic equations, in so far as it contains a nonlinear term with a space derivative of the first order, and a linear term with a second order space derivative multiplied by a factor which can be taken very small. Its properties with regard

↑Burgers at work, at home around the time that he published his main papers on what would become known as the Burgers equation. (courtesy of Burgers Archives / TU Delft)

to dimensions consequently are the same as those of the hydrodynamic equations. The solutions of [the equation] exhibit the property that regions make their appearance with very high values of | $\partial v / \partial y$ | ('steep fronts'), which can be considered as analogous to the regions of high dissipation appearing in the solutions of the hydrodynamic equations (regions of very high vorticity or shock waves)".

Yet in his farewell lecture in Delft in 1955 he had to remark: "Still the problem appeared exceptionally difficult – and only in 1953/54 I thought I had obtained a foundation, on which for this simplified system a statistical treatment is possible, though for the moment only for a so-called asymptotical case. The particularities which appeared were of an unexpected nature – and unfortunately I have to remark that no through going road can be found towards the real three-dimensional turbulence. At least, however, something has become visible of the particular properties of dissipative systems with a nonlinear differential equation."

Nowadays, we know that the 'flows' which are contained by the Burgers equation are essentially different from real turbulent flows, as Burgers gradually became to realize very well himself. As he remarked in his book The non-linear diffusion equation (published in 1974 when he was 79 years old): "The equation can be considered as referring to motions in an infinitely compressible medium, without pressure, and there is nothing in the system which deals with shear or with vortex motion". The Burgers equation had appeared to be too strong a simplification.

In recent years the Burgers equation seems to have been mainly used for testing and checking numerical solvers. On the other hand one is surprised by the broad scope of modern 'tools' which have been related to the Burgers equation: Markovian techniques, multifractals, wavelets, renormalization, solitons. Also the different fields of science in which the equation has been studied is impressive; apart from 'Burgers turbulence' one finds: nonlinear waves, the dynamics of growing interfaces (e.g., those of tumours), the dynamics of large-scale structures in the Universe, phase diffusion (e.g., neural signals), sedimentation, drainage, plasmas, waves in tubes, traffic flow, and acoustics. Further, the list of the various 'forms' of the equation is impressive: the forced Burgers equation, the noisy and the noiseless, the unstable, the discrete and the ultra-discrete, the viscid and inviscid, the generalized, the perturbed, the stochastic, the nonplanar, and the time-independent Burgers equation can all be found in recent literature.

## GROUNDWATER FLOWS

One particular field of fluid mechanics where several Dutch scientists and engineers have applied the theoretical approach has been the flow of groundwater, a field which belongs to hydrology. Rain water and water from canals and rivers creeps through sand (in the dunes) and the soil, which are porous media and in which capillary effects may be involved. In the 19th century the French engineer Darcy had already formulated a law describing this flow. In the Netherlands research on ground water flows had started at the end of the 19th century, both 'in the field' and in simple laboratories. Knowledge was required due to the growing demand for fresh tap water, which was partly made from rain water filtered by the dunes. After World War I the issue of ground water flows got even more interesting when a discussion arose about the consequences of the Zuiderzee project. It was Lorentz himself who had already studied the matter in 1913 and had shown, in the magazine De Ingenieur, how one could derive a Laplace equation for these kinds of flows. Van Iterson (see § 2.3.2) also wrote about it. The first hydrologist in The Netherlands who did a serious study on groundwater flow was Jan Versluys, whom we met as a researcher in Shell's Amsterdam laboratory (§ 3.3.2). In a paper of 1912 and in his PhD thesis of 1916 he studied the theoretical and experimental results which had been published worldwide and after comparing the validity of Poiseuille's Law and Darcy's Law, he concluded that the latter gives results for flow through porous media which are accurate enough. He also warned that at flow speeds groundwater flows can become turbulent and then neither Poiseuille nor Darcy are valid anymore.

Another early Dutch contribution was a theoretical study by J. Kooper in 1914 of steady flow around a well or circular polder in a leaky aquifer. Kooper arrived at a mathematical solution with Bessel functions, which proved to be very useful in groundwater exploration and management. In 1926 Burgers also published a paper in De Ingenieur on groundwater flow. By using the Laplace equation and applying a conformal transformation he produced steady and transient solutions for radial flow from partially penetrating canals into a low-lying confined aquifer. In the late 1920s experiments on groundwater flows were performed in Burgers' lab, probably related to the work which Gerrit de Glee (1897–1975) was doing for his PhD thesis in 1930,

under the guidance of Burgers. De Glee elaborated on the solution which Kooper had proposed and came up with a formula which is generally referred to as the De Glee formula.

Kooper and De Glee were both engineers with the National Institute for Drinking Water Supply (RID). This organisation, which was founded in 1913 to stimulate and support central drinking water supply (notably in the rural areas), remained the central agency for groundwater exploration and research for more than 50 years.

Jan Mazure (1899–1990), an engineer with the Zuiderzee Works (and in the 1960s the chairman of the Dutch Senate), saw that Burgers had not taken into account the resistance against upward leaking in his derivation. He therefore proposed a simplified flow pattern but more realistic hydrogeological schematisation by including a leakage factor. He assumed horizontal flow in the aquifer and vertical flow through the confining layer, which resulted in a simple exponential function for the steady-state hydraulic head distribution. With this formula the rise of the groundwater level at a certain distance from a river (or other water reservoir) could be predicted. It proved to simulate the actual observed situation rather well in the first Zuiderzee reclamation in the 1930s, the Wieringermeer polder, and has subsequently been widely used in the Netherlands. In 1937 Mazure gained his doctorate under the guidance of Burgers.

In the Dutch East Indies (Indonesia) Cor Vreedenburgh (1895–1936), professor of theoretical and applied hydrodynamics at the Technical University of Bandung, used mathematical tools to show, for the first time, an exact solution for the flow through and under an earth dam. Like Burgers, Vreedenburgh applied complex function theory to solve the two-dimensional Laplace equation. He also found solutions of the equations for flow through an anisotropic medium. In both these cases the permeability is no longer dependent on the coordinates but on the direction of the flow and therefore the Laplace equation is no longer valid. Vreedenburgh also did experimental work: he simulated two-dimensional groundwater flow in an electrolyte-tank on the basis of an analogy between Darcy's Law and Ohm's Law.

After the War theoretical contributions of Dutch researchers were still made but the rise of numerical simulations also had its own influence. In 1949 Henri Brinkman (1908–1961) published an extension of Darcy's law which can be used to take account of the transitional flow between boundaries. It became known as the Darcy-Brinkman equation and still regularly appears in papers on flow in porous media and on other topics.

When Gerard de Josselin de Jong (1915–2012) became professor of applied mechanics in Delft in 1961, he had already been in California for some time where he had worked on the modelling of the transport of a pollutant in the flow of groundwater. He argued that the mechanism of dispersion in

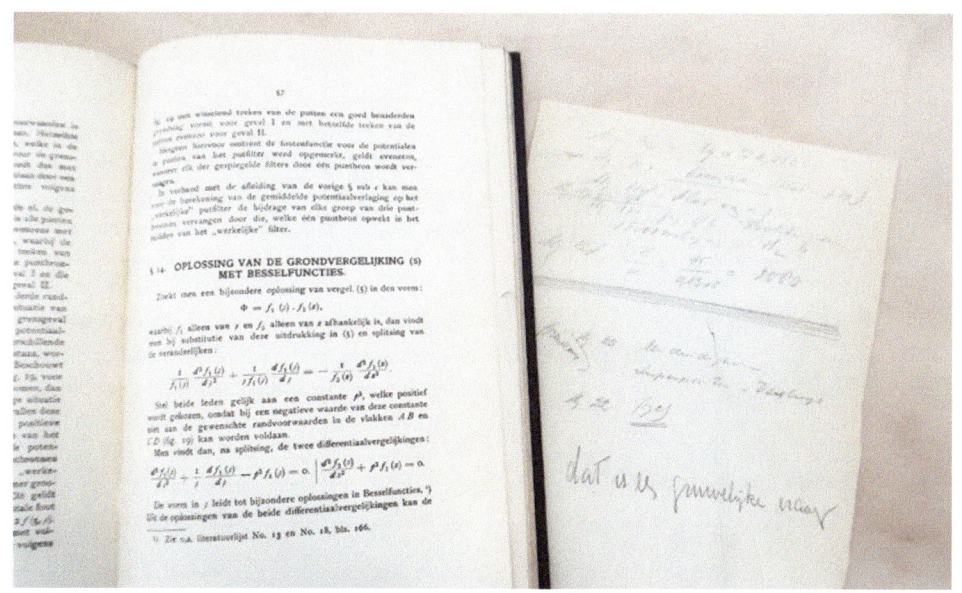

←A page from the PhD thesis of De Glee (1930) and a note which was found in the copy present in the Burgers Archives. The handwriting at the bottom definitely is that of Jan Burgers and (in translation) it says: "that is a gruesome question". (courtesy of Burgers Archives / TU Delft)

the direction of flow differs from the dispersion in the lateral direction. This leads to a much larger effect of dispersion in the flow direction than in the transverse direction. He validated his theory by careful experiments. De Josselin de Jong also worked on the vortical motions along the boundary between fresh and salt water due to opponent flows along this boundary.

Analysis of the drainage process led in 1958 to the transient formulas of De Zeeuw & Hellinga and Kraijenhoff van de Leur (see also § 4.1.4), in which the transformation of rainfall into groundwater discharge is characterised by a reservoir factor. This factor includes the hydrological properties of the subsurface and the geometry of the drainage system. These formulas formed the foundation for the design of a drainage scheme on the basis of the required groundwater depth. The optimal groundwater depth was subsequently determined − for various crops and soils − by study of the flow of water in the unsaturated zone, and notably of the water supply by capillary rise.

Ernst (1962) developed solutions for non-linear flow processes towards drains, thereby including the geometry of the drainage channels and the influence of heterogeneous subsoil. Both the solutions of Hooghoudt (see § 6.2.6) and Ernst have been applied worldwide. In the 1970s a group of hydrologists came up with an expression (sink term) for water uptake by roots within the continuity equation for unsaturated groundwater flow. This equation formed the basis for a general numerical soil-water-atmosphere-plant (SWAP) model, to describe transient water and solute flow in heterogeneous soil-root systems. This model is also used worldwide.

## PHYSICISTS, ENGINEERS, AND MATHEMATICIANS

The mathematical approach to flow phenomena was, one could say, simply the most logical approach for physicists like Burgers and Broer. They were trained in it (Broer originally in another branch of physics) and they were good at it. Most engineers were less well-trained in this respect but naturally there were exceptions, especially among those who had chosen the more 'physical' directions, like solid and fluid mechanics at the Department of Mechanical Engineering in Delft. These were also the directions with the least students and with the stamp 'difficult' on them. Hinze and Van Wijngaarden, students of Burgers, are among these exceptions.

From the 1950s a 'love' of mathematics could especially be found among the Delft aerodynamicists. At the NLR many of them could find a lot of enjoyment in the theoretical groups. The Dutch poet, novelist, and essayist Gerrit Krol (1934−2013) had been trained as a mathematician and after his graduation in Amsterdam, he was looking for a job. In his book In dienst van de Koninklijke (about the time he worked for Shell), he wrote about this: "I had applied for a job at the Luchtvaartlaboratorium [the NLL; FA] in Amsterdam, had visited it and I had seen that in the shortest possible time I would become a specialist in solving differential equations there." Krol never took the job he was offered. But people like Zandbergen and Van Spiegel (see chapter 4) did.

In other countries several 'pure mathematicians' had discovered fluid mechanics from about 1900. (One example is Otto Blumenthal from Aachen who fled to The Netherlands in the late 1930s where Burgers and others tried to help him. He died in Theresienstadt in 1944.) Among the Dutch mathematicians examples of an interest in flow theory are hard to find. Only from the 1950s did the situation start to change which, we can suppose, was largely due to the start of the training of 'mathematical engineers' (or 'applied mathematicians') in the TH in Delft. One of the most important pacemakers of this development was Timman (see below).

A new approach to fluid mechanics, especially turbulence,

←When Jaap Steketee retired as professor of theoretical aerodynamics in Delft in 1992, he must have been one of the last in fluid mechanics to have never touched a computer. Steketee was known for his very clear lectures and for his ability to apply mathematical techniques in aerodynamics. He had done some experimental work when still a student but had never liked that. (reprinted from Bakker et al (1992), with permission from IOS Press)

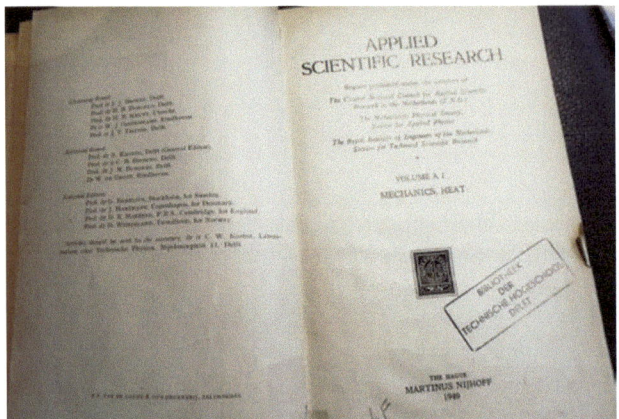

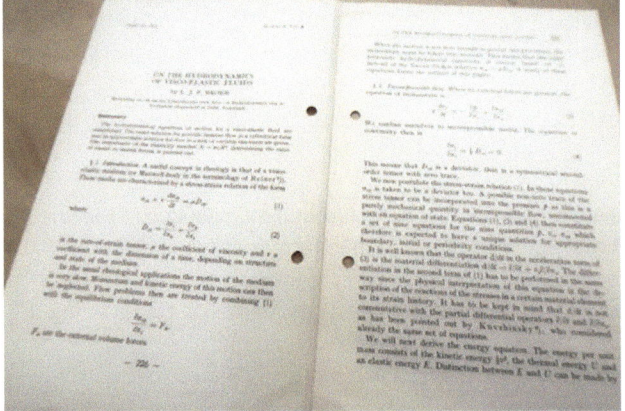

↑↗In 1947 some Dutch physicists thought it was time to start a new journal, published in their own country, in which articles could find a place which were hard to get published in the existing Dutch journal Physica, the journal of the theoretical physicists. The first volume of Applied Scientific Research would appear in 1949. Initially, the editors welcomed contributions from a broad range of physical fields but from the beginning it was also clear that fluid mechanics would be one of the prominent fields. In 1990 the editorial board, consisting of Nieuwstadt, Ooms, and Van Wijngaarden, decided to change the subtitle into Applications of Fluid Dynamics and in 1998 the title itself was changed to Flow, Turbulence and Combustion. Remarkably, Burgers himself would never publish in this journal. But Broer did and in volume 6, e.g., he wrote down the equations for a visco-elastic fluid, which – in hindsight - seems a rather bold thing to do in the 1950s. Like Burgers, Broer was much at ease with deriving and manipulating mathematical equations. (original publisher: Nijhoff, The Hague / courtesy of Springer)

came in the 1970s and found its origin in theories developed by mathematicians. The Burgers equation has been called "one of the simplest model equations which can exhibit spatiotemporal chaos". The 'chaos theory' became very popular in fluid mechanics in the 1980s (also with the general public), when it already had quite a long history. In 1971 a paper was published which became well-known, entitled 'On the nature of turbulence', written by Ruelle and Takens. Floris Takens (1940–2010) was a Dutch mathematician and professor in Groningen and became internationally known for his contributions to the theory of chaotic dynamical systems. Together with the Frenchman David Ruelle, he predicted that fluid turbulence could develop through a strange attractor, a term they coined, as opposed to the then-prevailing theory of accretion of modes. The prediction

was later confirmed by experiment.

Takens also established the result now known as the Takens' theorem, which shows how to reconstruct a dynamical system from an observed time-series. This theorem has been applied to some multiphase systems used in the process industry from around 2000 by the group of Cor van den Bleek (1943), then professor of chemical reactor engineering in Delft. This monitoring method used a characteristic process variable, e.g., pressure, measured at high frequency. The obtained time-series was transformed into a so-called attractor, representing the successive states of the system during its evolution in time. The consecutive attractors obtained during the operation of the process were compared with a reference attractor reflecting the desired behaviour.

# R. TIMMAN (1917–1975)

Reinier (or Rein) Timman studied mathematics and physics and found a job at the Fokker airplane factory in Amsterdam in 1939. During the war he put himself to study aerodynamics by reading all volumes of the Zeitschrift für Angewandte Mathematik und Mechanik and of Luftfahrtforschung. He discovered that some German scientist had been working on the vibrations of two-dimensional wings in flows of incompressible air (flutter). If these vibrations are not damped, then the wings can break. Timman succeeded in tackling this problem for flows of compressible air and managed to hide his results away from the Germans.

After the war Timman got his PhD in Delft where one of his supervisors was Jan Burgers. He left Fokker and got a position at the NLL where his expertise in the mathematical approach of flutter was very welcome. However, Timman's formulas had to be translated into practical results and to this end a huge amount of calculating work had to be done. This was done at the still young Mathematisch Centrum (Mathematical Centre, MC) in Amsterdam. It took several years before the MC could provide the calculated values of the aerodynamic coefficients which Timman and his team needed. But then, a terrible thing happened. An American scientist showed that the coefficients couldn't be right! The NLL tried to refute the criticism but had to admit that one of the formulas used showed an omission... After this was corrected, the recalculated air forces appeared to be correct.

Timman stayed at the NLL for six years and produced a lot of useful papers. He got involved in boundary layers and was able to extend the calculations to three-dimensional flows. He showed how to use tensors to describe the flow around a body. He considered the effect of two parallel walls on a vibrating wing profile and thus was able to provide the experimenters a method to correct their measurements from wind tunnel experiments (where the influence of the walls cannot always be neglected). Thanks to Timman the NLL got an excellent reputation, both national and international, with regard to non-stationary aerodynamics.

Despite Timman's success at the NLL, he chose to become a professor in 'pure and applied mathematics' at the Delft University, in 1952. From the first day in Delft he was aiming at the foundation of a new field of study in Delft, and of a new type of training for engineers; i.e., engineers with such an extensive knowledge of mathematical techniques would be able to tackle any problem from engineering practice. In 1956 the Ministry of Education finally gave Delft permission to start the education of 'mathematical engineers'.

Timman remained advisor at the NLL after he had left for Delft. In 1955 he also became advisor at the NSP where at that time hardly any mathematical activity took place.

Scientists at the NSP started to do mathematical research on ship propellers and on 'slender bodies'. In his group in Delft the interest in hydrodynamics also grew, resulting e.g., in a thesis on cavitation. The WL also approached Timman and his co-workers for advice on the modelling of complicated flow phenomena.

Timman's reputation was worldwide, he even became advisor of the David Tayler Model Basin in the USA which did work for the American navy. In 1957 Van der Maas gave an extensive lecture before the Royal Aeronautical Society in London. One of the attendants there, praised one of the mathematical contributions by Timman to fluid mechanics: "[I would like to mention] Timman's brilliant theoretical discovery in the field of two-dimensional oscillating derivatives in high subsonic range. This beautiful analytical solution, by means of Mathieu functions, ranked high in aerodynamic theory, and was something of a shock for many predecessors who, despairing of the possibility of an algebraic solution, went the defeatist way of relying on brute computing force (the war which threatened to become a time-honoured escapist way in many parts, thanks to the monstrous development of electronic brains!). The solution was severely attacked because of some initial discrepancies between old and new arithmetical results but, as often happened, the errors were found to be computational only, and the new theory emerged victorious, to the great delight of 'algebraists'."

In 1975 Rein Timman gave a keynote address at the First International Conference on Numerical Ship Hydrodynamics in the USA. A few weeks later he died. One year earlier his son Jan had become a Grandmaster and had started an impressive career in chess.

↑Timman at the far end of the tabel on the left, in Paris in 1972. With Aad Hermans (whois sitting opposite Timman), he took part in the 9th Symposium on Naval Hydrodynamics. (courtesy of Aad Hermans)

| Name | Standard symbol | Definition |
|---|---|---|
| Archimedes number | Ar | $\mathrm{Ar} = \dfrac{gL^3 \rho_\ell (\rho - \rho_\ell)}{\mu^2}$ |
| Atwood number | A | $A = \dfrac{\rho_1 - \rho_2}{\rho_1 + \rho_2}$ |
| Bejan number (fluid mechanics) | Be | $\mathrm{Be} = \dfrac{\Delta P L^2}{\mu \alpha}$ |
| Bingham number | Bm | $\mathrm{Bm} = \dfrac{\tau_y L}{\mu V}$ |
| Biot number | Bi | $\mathrm{Bi} = \dfrac{hL_C}{k_b}$ |
| Blake number | Bl or B | $B = \dfrac{u \rho}{\mu (1 - \epsilon) D}$ |
| Bond number | Bo | $\mathrm{Bo} = \dfrac{\rho a L^2}{\gamma}$ |
| Brinkman number | Br | $\mathrm{Br} = \dfrac{\mu U^2}{\kappa (T_w - T_0)}$ |
| Brownell–Katz number | $N_{BK}$ | $N_{BK} = \dfrac{u \mu}{k_{rw} \sigma}$ |
| Capillary number | Ca | $\mathrm{Ca} = \dfrac{\mu V}{\gamma}$ |
| Chandrasekhar number | C | $C = \dfrac{B^2 L^2}{\mu_0 \mu D_M}$ |
| Colburn J factors | $J_M, J_h, J_D$ | |
| Damköhler number | Da | $\mathrm{Da} = k \tau$ |
| Darcy friction factor | $C_f$ or $f_D$ | |
| Dean number | D | $D = \dfrac{\rho V d}{\mu} \left( \dfrac{d}{2R} \right)^{1/2}$ |
| Deborah number | De | $\mathrm{De} = \dfrac{t_c}{t_p}$ |
| Drag coefficient | $c_d$ | $c_d = \dfrac{2 F_d}{\rho v^2 A}$, |

↑All experimenters in fluid mechanics know: to make your model experiments relevant, you should keep the dimensionless numbers in the laboratory equal to those in the real world. In the past century many dimensionless numbers have been given a name. One of the most famous is the Reynolds number (Re). Only very few have been named after Dutchmen. We mention the Brinkman number (Br), named after the mathematician and physicist Henri Brinkman (1908–1961) which is related to heat conduction from a wall to a flowing viscous fluid, commonly used in polymer process-ing. Unfortunately, one can completely ignore Br since it can be replaced by Pr · Ec where Pr is the Prandtl Number and Ec is the Eckert Number, both named after non-Dutch scientists... (photo from Wikipedia: en.wikipedia.org/wiki/Dimensionless_numbers_in_fluid_mechanics [2018])

↑During Burgers' experiments on flows behind objects in the small towing tank, also stereo photos were made. These were taken by a camera moving along with the objects. (courtesy of Burgers Archives / TU Delft)

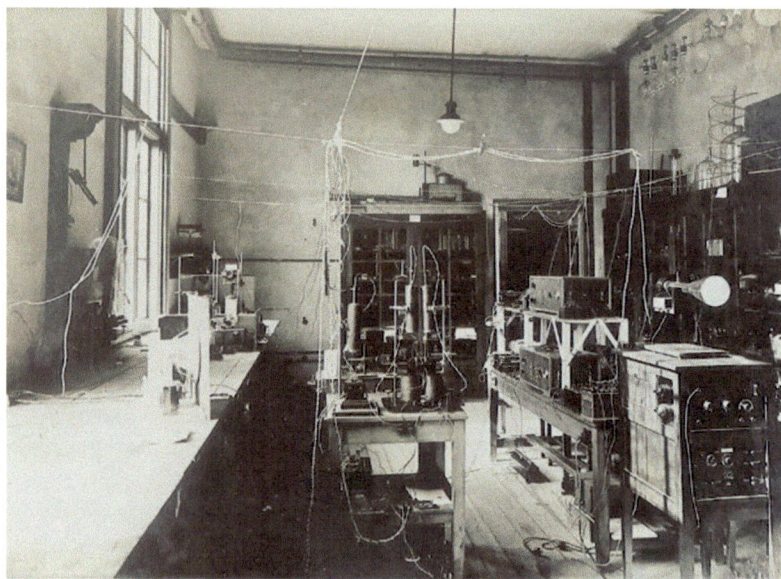

↑In 1930 all interesting places in Burgers' laboratory were photographed. This must have been the place where the experimentalists built and tested their equipment. We may suppose that some of the 'electronics' is related to Ziegler's work on amplifiers for the hot-wire technique. (courtesy of Burgers Archives / TU Delft)

# 6.2 EXPERIMENTAL APPROACHES

"I would like to say some words about the science of liquid and air flows. Although here the foundations were deter-mined a considerable time ago, the practice of these fields is somewhat different from that in mechanics because it is so much more difficult to derive results along a mathemat-ical way. Therefore one has to resort to experiments". Thus spoke (in Dutch) the well-known physicist Hendrik Casimir in his inaugural lecture as special professor at the University of Leiden in 1939. Casimir didn't consider a third way to do research in fluid mechanics, the numerical approach (see § 6.3), which is not so strange since in the Netherlands this

kind of research was hardly done in the 1930s. It was only in the 1960s, due to the rise of the digital computer, that Casimir's words would become outdated.

## • 6.2.1 MEASUREMENT AND VISUALIZATION TECHNIQUES

MEASURING AND VISUALIZING IN BURGERS' LABORATORY UP TO 1955

The very first experiments done in the Laboratory for Aero-dynamics and Hydrodynamics were aimed at visualizing and

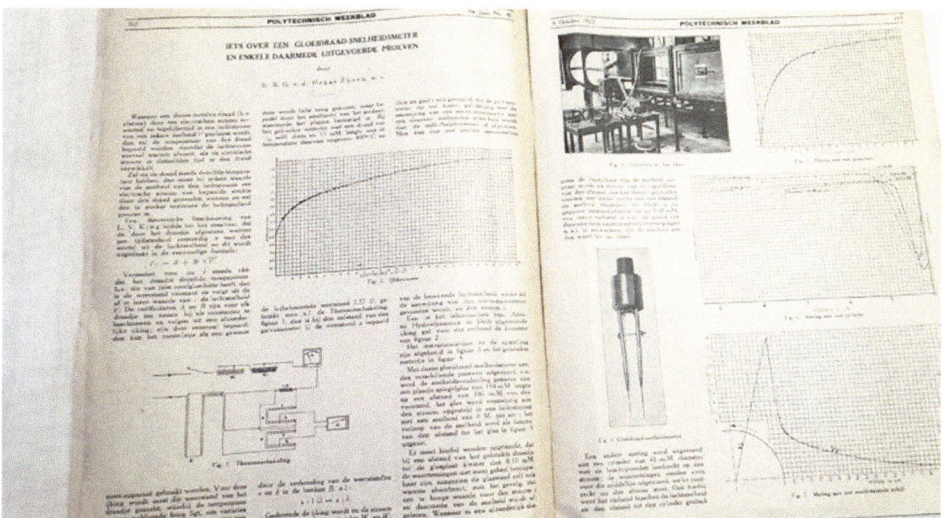

← In 1922 Van der Hegge Zijnen wrote about the still unknown hot-wire anemometer in the *Polytechnisch Weekblad*, a journal for engineers. He also explained how this device could be calibrated. To capture the fluctuating signal of his hot wires, Van der Hegge Zijnen used photo paper on which the signal could be depicted. (courtesy of Burgers Archives / TU Delft)

↑ Van der Hegge Zijnen turned out to be a clever experimentalist. In the 1920s he also developed a special Pitot probe with five holes. With this the static pressure and both the value and the direction of the flow velocity could be measured. This probe has also been used for an investigation into a problem in one of the huge ventilation systems of the State Mines (see § 3.3.1). It detected a 'disturbing vortex' in one of the channels which could easily be removed by placing a small partition. Here we see the special probe on the left, together with other Pitot tubes which are conserved in a showcase at the section of Fluid Mechanics in Delft, the successor of Burgers' laboratory. (courtesy of Burgers Archives / TU Delft)

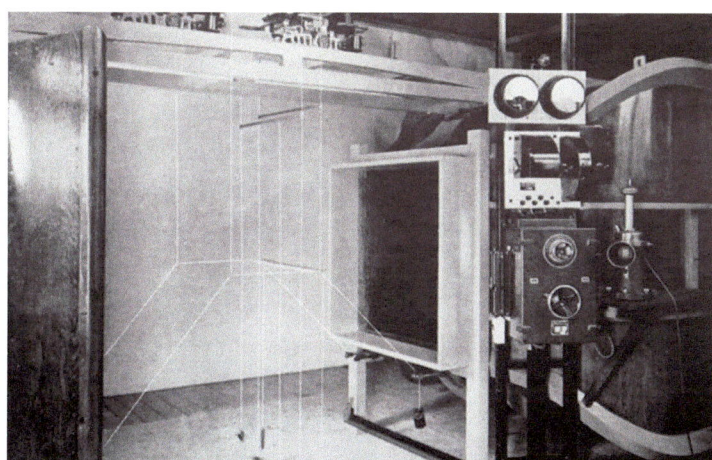

↑ Besides flow velocity and pressure, sometimes forces on objects induced by air flows had to be measured in Delft. Since up to 1953 Burgers' Laboratory was the only place of the TH where aeronautical research could be done, research on aerofoils had to be done in one of the wind tunnels of the Laboratory for Aerodynamics and Hydrodynamics. To measure the forces on a wing profile an ingenious construction of weights and balances was built. This same mechanism was also used to measure the characteristics of several models of windmill sails (see § 3.4.2). (from: Havinga (1959))

observing flows around, and especially behind, objects like cylinders, flat plates, and aerofoils (or airfoils). These objects were towed in the water tunnel/towing tank of the laboratory (see § 6.2.4). The flows were made visible by means of aluminium powder floating on the water.

The idea of the boundary layer had been introduced by Prandtl in 1904 but up until about 1920 nobody was really able to do measurements on it. In 1921 Burgers got acquainted with the research that Von Kármán was intending to do in Aachen and this inspired him to also take up exper-

iments on boundary layers. In the early 1920s there were only three institutes, Göttingen, Aachen, and Delft, that were interested in this subject. In the period from 1904 to 1924 in scientific journals only about twenty papers on boundary layer problems were published.

Burgers read about a new measurement technique in which tiny metal (e.g., platinum) wires were used which were connected to a battery (the resistance was kept constant so that the change in current strength could serve as an indicator of the air speed). Attempts were made in Delft to use these wires in the towing tank but these appeared to have been

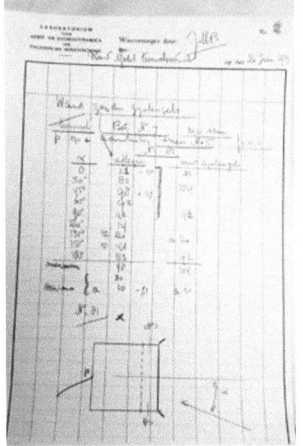

←To write down measurements Burgers' Laboratory had its own special forms. This document from the Burgers Archives shows that by 1939 Burgers himself did some measurements in one of the wind tunnels. (courtesy of Burgers Archives/ TU Delft)

↑Flow visualization in the 1930s. Around 1936 a temporary 'shed' was built close to Burgers' Laboratory for experiments on the ventilation channels of the Maastunnel (see § 3.4.3). Parts of the model were made transparent and strings were placed inside which would follow the air flows. On the left one sees one of the many Betz precision manometers which were then used in the laboratory. Outside the building a model of the complete ventilation building was tested using smoke. (courtesy of Burgers Archives / TU Delft)

not very successful. In 1923 Van der Hegge Zijnen, Burgers' main research assistant for many years, started the now classical experimental work on the flat plate boundary layer in one of the wind tunnels of the laboratory. He improved the hot-wire technique and became the first who was able to measure velocities as close as 0.05 mm from the wall. He could confirm a result which had been predicted by Prandtl and Von Kármán some years earlier, namely that the mean velocity in a turbulent boundary layer increases as y1/7 with y the distance from the wall of a flat plate. In 1924 Van der Hegge Zijnen finished his PhD thesis on this research.

Subsequently the hot-wire method was increasingly used in the Laboratory to measure turbulent fluctuations and correlations. Burgers invented some new hot-wire anemometers himself, one of them with two parallel wires. It became clear that this equipment required more knowledge of electronics than was available among the staff. It was a great help when, in around 1928, another assistant could be appointed, Marc Ziegler, who had an electronics background (and would leave Delft after only a few years for Philips). One of the interesting results which was obtained has been described by Burgers as follows:
"When making oscillographic records of the velocity fluctuations in the boundary layer along a glass plate, he (Ziegler)

again found that the boundary layer can be steady and laminar over a certain distance, while in the region of transition this laminar flow appeared to be interrupted at irregular intervals by short periods of complete turbulence. These periods were found to grow in duration and number as one goes downstream. Ziegler thus observed the intermittency of incipient turbulence."

Despite many broken wires and a growing number of alternate, and sometimes more convenient, techniques, the hot-wire technique kept its place in the Delft laboratory until well into the 1990s. In the 1950s the equipment had already become so shock-proof that it was used for boundary layer measurements aboard the Koolhoven FK 43 plane of the Department of Aeronautical Engineering. The principle changed from 'constant current' to 'constant temperature' which allowed a much higher frequency response, much needed for research on turbulent flows. Another improvement was the replacement of a single wire by two wires arranged in an X. This made it possible to measure the so-called Reynolds stress in the flow (the friction caused by the turbulence). The last probe made in the Laboratory even had four wires and was fabricated by a specialist in a special workshop under a microscope in 1994.

## WATER-RELATED MEASUREMENTS ELSEWHERE

←A typical situation which one could encounter for decades in the Waterloopkundig Laboratorium in Delft. This photo was taken in the 1950s, when the use of self-registering instruments was still not common practice. For most measurements the people at the WL sometimes had to be inventive since no standard equipment was available to do the job. It is known that sometimes Meccano elements and letter scales were used. (courtesy of Deltares)

Fig. 130. Bodemstroommeter voor stroomen op 0,15 en 0,50 m + bodem.

→In the 1930s Rijkswaterstaat had started to do research on the Delta area. Some of their engineers started to perform measurements, e.g., on the speed and direction of the water flows near the bottoms of tidal inlets and on the movements of sediments. To this end they had to develop specific measuring instruments which could be used on special measuring ships. On the tidal movements near the Dutch coast, still hardly anything was known and therefore RWS decided to install self-registering tide measuring devices. For the investigations on sediments, RWS could make use of this invention by J.J. Canter Cremers, an engineer who had been involved in the state committee that had studied the consequences of a continuation of the deepening of the Nieuwe Waterweg (see § 2.1.1). He designed a 'zandvanger' (sand catcher). This page from Van Veen's description of the measurements at sea also shows a 'bottom flow meter'. (from: Van Veen (1936))

Fig. 132. Het ophalen van den zandvanger.

↑Around 1960 the use of electronic equipment had become common in the towing tanks of the Shipbuilding Laboratory of the TH Delft. But some handwriting was still necessary (note the pipe). Ten years later, at the NSP in Wageningen, the researchers seem to have been definitively liberated from pen and paper and the equipment looks even more impressive. (courtesy of TU Delft / photo by Fotografishe Dienst via Beeldarchief / CC BY; courtesy of MARIN)

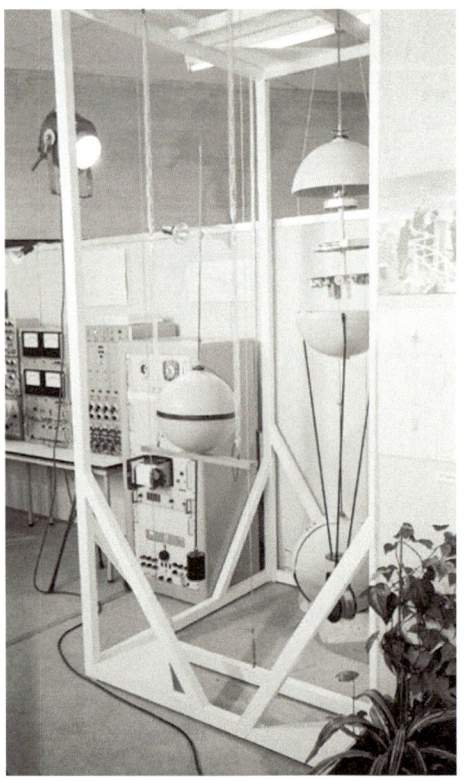

←Measurements outside the laboratory sometimes required the invention of new and robust measuring devices. These buoys could be used for measuring wave heights at sea. According to a description given by a TH Delft newsletter in 1967 the procedure was as follows: "At the start of the measurements a buoy, equipped with a accelerometer and a radio transmitter, is thrown into the sea. The measured acceleration signal – frequency modulated – is emitted and received on board, demodulated and integrated in time twice, to find the wave height." (courtesy of TU Delft / photo by Fotografishe Dienst via Beeldarchief / CC BY)

## AIR-RELATED MEASUREMENTS ELSEWHERE

↑The measurements like those with the balances delivered a stream of data which had to be 'processed'. Up until the 1950s calculations on these data were done by hand at the Uitwerkdienst (Data Processing Service) of the NLL. This photo of about 1946 suggests that this Service was completely run by women. (courtesy of Stichting Behoud Erfgoed NLR)

↑From the 1980s full-scale wind turbines were tested in the dunes at the Energy Centre of the Netherlands ECN in Petten. In 1993 Gustave Corten invented the so-called stall flag method as a PhD student in Delft. Corten continued the development of the method at ECN. The method is still the only option to visualize the dynamic stalling behaviour of full-scale wind turbines. The stall flag is a sticker-like indicator with a hinged flap covering a reflector. When during the rotation of the blades separation occurs at a certain spot, the flow will be reversed and the flap will turn over. An observer in the field with a lamp can see the flagging separation. The method was used to study dozens of turbines in several countries. Corten's company applies the method to detect and solve premature stalling problemsn by the application of the so-called vortex generators and thus increasing the energy yield. (courtesy of Gustave Corten / CortEnergy)

↑In the world of dredging the equipment is always robust. This is a combined speed and concentration meter, developed by Krohne and IHC and exhibited at the National Dredging Museum in Sliedrecht. With this meter the 'zuigbaas' (suction boss) can be constantly informed about the precise composition and flow speed of the slurry which is taken from the bottom of the waterway. The working is based on principles from electromagnetics and radioactivity.

←The design of airplanes involves the testing of aerodynamical properties. To this end models and wind tunnels were used almost from the start of aviation. The Fokker F.II was tested upside down in the first wind tunnel of the RSL around 1920. The strut was connected to the so-called Eiffel balance, mounted above the test section. The picture shows a testing session around 1945, in the small low-speed tunnel of the NLL (tunnel no. 4). The model is suspended on wires attached to an external 'overhead' balance. (courtesy of Stichting Behoud Erfgoed NLR)

↑In the 1950s the operators' areas of wind tunnels became more 'professional' and resembled that of industrial process plants. This is the first wind tunnel of the Department of Aeronautical Engineering in Delft, where in 1963 a group of visitors gets an explanation of the tunnel by professor Van der Maas himself (leaning on the console). Note how all guests are focussed on the small printing device. Data from the wind tunnel tests could now be recorded automatically. (courtesy of TU Delft/ photo by Fotografishe Dienst via Beeldarchief / CC BY)

↑In the world of aeronautical research, one sometimes needs a real airplane to do full scale measurements for other real airplanes. This Cessna Citation was a 'laboratory plane' of the NLR around 2000. For a large international project called S-Wake, it was equipped with a 'nose boom' on which flow vane sensors could measure the wake behind planes. If two planes at an airport take off after each other, there is a chance that the vortex wake of the first one will cause a crash of the second one. It was still unknown at that time which distances between planes could be regarded as safe. (courtesy of Stichting Behoud Erfgoed NLR)

↑To determine the locations of transition spots where the laminar boundary flow had turned into a turbulent flow, the NLL used the so-called China-clay method. Here, in 1957, the method has been applied to the rotor blade of a helicopter. The dark parts indicate a laminar flow. (courtesy of Stichting Behoud Erfgoed NLR)

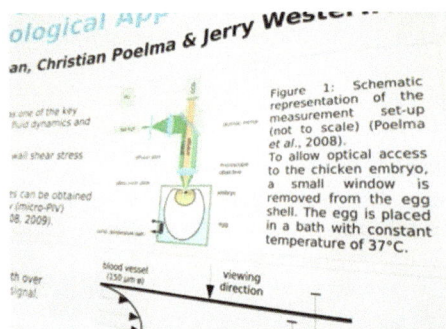

↑ Two applications of LDA. At the IFRF laboratory in IJmuiden (see also Ch 7) LDA has been used to measure the (turbulent) flows related to flames and burners.(courtesy of IFRF) At Shell LDA was used from the 1970s but only with the arrival of powerful computers in the 1980s this technique became really fruitful. (courtesy of Shell / Peter Veenstra)

↑ Westerweel and Poelma have been involved in several PIV-related investigations. One of their first posters on their results can probably be claimed as the only scientific presentation poster which is completely wordless. It shows a flow field of 40 x 40 mm showing details of 2 micrometers. The other photo shows a detail of a poster on blood flow measurements. PIV is by far the most suitable method for this since other methods, like LDA, imply a temporal resolution which is longer than typical heart rates. The project involving a chicken embryo was a project in cooperation with medical groups in Leiden and Rotterdam. In a research project which started around 2005 PIV was used to investigate whether extra-embryonic flow influences heart development. This was based on the speculation that heart development is linked to the wall shear stress patterns within the forming heart. In vitro studies have shown that shear stress can modulate gene expression.

↑ Also in research on airplane models PIV has greatly improved the possibilities of obtaining data. Around 1985, in the wind tunnel of the NLR, a probe had to be moved along the model. Today the flow field can be measured much quicker by using a green laser sheet and PIV technology. (courtesy of Stichting Behoud Erfgoed NLR)

## THE LASER ERA

Soon after lasers became reliable and (more or less) affordable in the early 1960s, they became used for many purposes. One of them was the measurement of fluid flows. The laser entered the laboratories and warning signs ('laser on') and protective sheets of black and blue 'agricultural plastic' appeared everywhere.

In 1964 Laser Doppler Anemometry or Velocimetry (LDA, LDV) was introduced to the scientific and industrial world. In LDA the Doppler-shift of light scattered by a small particle that moves with the flow is determined. This Doppler-shift provides a measure of the velocity of the particle, and therefore the flow velocity. The main advantage of the technique over conventional measuring techniques, such as hot-wire anemometry and pressure probes, is that it does not require a physical probe in the flow, i.e., it is a non-intrusive technique. Therefore, the flow is not disturbed during a measurement.

It is not known when LDA was used for the first time in the Netherlands but in the Kramers Laboratory in Delft the technique was used from the early 1970s. At the Technisch Physische Dienst TNO-TH (known as TPD) LDA was tested and developed in 1971. Soon LDA would enter almost all fluid mechanics laboratories, e.g., the flame research facility of the group of Hoogendoorn in Delft.

But LDA and hot-wires are one-point measurement techniques, and therefore not able to reveal the instantaneous spatial structure of a flow. For the detection of coherent structures in turbulent flows, and for other phenomena, a more sophisticated technique was needed. This became Particle Image Velocimetry, or PIV. In this technique a flow is seeded with small particles. These are illuminated in a plane by means of a laser light sheet and are recorded by a camera. By taking two exposures of the same image separated by a small time fraction, one records for each particle two positions displaced by a small distance. By means of a correlation analysis one can measure this distance and after dividing this by the time interval one obtains the velocity vector field in the light plane.

PIV had already been developed in the USA in the 1980s but was then in an 'analogue' and therefore time-consuming version. Around 1990 Jerry Westerweel in Delft, under the guidance of Nieuwstadt, started to develop a digital version of PIV, independently of the American researchers working on the same topic. In his PhD thesis of 1993 Westerweel was able to show that Digital PIV (DPIV) was a fast and reliable technique which could produce results of "acceptable accuracy".

From then on, PIV started to conquer the world of fluid mechanics and soon several variants of PIV appeared. For Stereoscopic PIV (SPIV) two cameras were used. (SPIV was used e.g., by the group of Clercx and Van Heijst for experiments on rotating Rayleigh-Bénard flows; see § 5.1.3). Then came the quest for PIV variants with which 3D patterns could be found in flows. One of these variants was Holographic PIV, another was Tomographic PIV. An important early contribution to Tomo-PIV came from Gerrit Elsinga in 2006 (for which he was awarded the Leen van Wijngaarden Prize in 2012). The technique was also further developed by Westerweel and Christian Poelma.

Since 2000 PIV has been used for many purposes: to detect coherent structures in the compressible wake of a 2D model in a supersonic freestream; to measure blood flow in a chicken embryo; to study cavitation on hydrofoils; to investigate the flow above rippled sand beds; etc.

For some flows, PIV does not work. One of the difficulties with experiments in multiphase flows is the opaque nature of most multiphase systems. Consequently, laser-based techniques are of limited use. However, matter is to a certain extent transparent to X-rays. The group of Rob Mudde in Delft (professor of Multiphase Flow at the Department of Chemical Engineering) has used a fast X-ray CT scanner. With this device they could reconstruct the spatial distribution of the various phases in two parallel cross sections of a model reactor. Poelma (also professor of Multiphase Flow, but at the Department of Mechanical, Maritime and Materials Engineering) is working on 'tomographic reconstruction using X-rays'. Other non-PIV techniques have also become more or less common in fluid mechanics. The high-speed camera has proved itself to be very useful in the investigation of the behaviour of droplets and bubbles (e.g., in the group of Lohse in Twente). MRI has been used for flow measurement by professor Daniel Bonn of the University of Amsterdam. The Applied Molecular Physics group of the University of Nijmegen is working on APART: Air Photolysis and Recombination Tracking. In this latter measuring technique velocity information is obtained by following tagged molecules in the flow, so it does not require particle seeding, as does the PIV technique. (The University of Nijmegen has been the only general university with a chair in fluid mechanics, at the Physics Department.)

## • 6.2.2 WIND TUNNELS

The 'general public' will associate fluid mechanics most of all with wind tunnels. The first wind tunnels were built in the last decades of the 19th century and there was a tremendous increase in their number, performance, and variety, during the 20th century, which continues even today. On the other hand, wind tunnels are disappearing from fluid mechanics, mainly due to the rise of CFD. For instance, at Shell's KSLA the wind tunnel was removed when the computer models become 'good enough' to investigate and evaluate the fluid flows relevant for the company. Some (remarkable) Dutch wind tunnels are treated in this section.

## EVERYTHING FLOWS, BUT HOW DO YOU MEASURE THAT?
JOOST GROEN, VSL B.V.

In the current-day world, particularly in western society, fresh water flows from our kitchen or bathroom tap when we open it, we can get petrol from a station close to where we are, and high-pressure natural gas flows through massive pipelines over large distances, often crossing national borders. It's as simple as that. We also pay for all this based on measured data, and we assume that the amount of water, petrol, or gas that we pay for is correct. Why is that? Indeed, we take it as read that the data supplied by the flow meters directly relate to the real amounts. In other words, that the meters are calibrated.

How should you calibrate a flow meter? A straightforward way is to use scales, or a flask of known volume and a timer, but you could also use another flow meter. However, such a calibration only holds water (so to say) if those devices are actually more accurate, and are themselves also calibrated. Clearly, that requires other, even more accurate, calibrated equipment, and so on. This chain can continue for a while, but soon it runs out of links. At that point you have arrived at the most accurate, highest measuring standards that exist. These are guarded, maintained, and improved by the National Metrology Institutes (NMIs) of the world.

Metrology is the science of measurement. Measuring properly and accurately, particularly at the highest level, is certainly a craft, and by many even considered as an art. In any case it is not straightforward. The NMI of the Netherlands is VSL, or Van Swinden Laboratory. Jean Henri van Swinden was a Dutch scientist who was one of the people who formed the basis of the current Système International d'Unités, better known by its abbreviation as the SI. In 1799, Van Swinden handed the original meter (a platinum rod) to the French government. In 1816 the Netherlands was the first country to have a Metrology Act and in 1929 became a member of the Metre Convention. Since 1937, VSL and its predecessors have been responsible for the Dutch measurement standards, by request of the Dutch government.

Just a small clarification: the flow meters in your home are not really calibrated, but they are verified. This means that they are checked against a (calibrated) standard and if the deviation of the meter is below a certain value (the maximum permissible error), it receives the "OK" stamp. Calibration is a more complex process, where for several values of the flow range of a meter, the deviation with the standard is determined and documented. What's more, for each of these values, the uncertainty of that measurement is assessed. Uncertainty is a key parameter in everything VSL does. The lower the uncertainty of a given measurement, the more accurate the measurement. A proper uncertainty determination relies on an unbroken chain of well-documented calibrations from the measurement under consideration all the way to the highest measurement standards. If that is the case, the measurement is said to be traceable to the SI. Without traceability, a measurement result formally means nothing. The determination of the uncertainty is almost a science unto itself, often involving complex calculations. Particularly for industrial applications, uncertainty can be difficult to determine. A flow meter is more often than not used at a different temperature and pressure as well as for a different medium than used in calibration. These differences are all accounted for in extra uncertainty terms, often after long debates. While a value of 0.25% uncertainty, let alone a difference between 0.24 and 0.25% uncertainty might seem small or even insignificant, if a very large amount of, for instance, natural gas flows through the meter (which happens with the meters at the borders of our country), the amount of "uncertain" gas can be huge. When the discussion turns fiscal, in other words when the measurements are used to charge for the fluid measured, the impact can be massive.

Over the last century developments in liquid and gas flow measurement have been immense. Current ubiquitous mechanical flow meters for instance involve bluff bodies that generate Von Karman vortex shedding, turbines, and rotors. Flow meters using temperature differences, an electromagnetic field, or ultrasound, do not physically reach into the flow. Coriolis flow meters measure mass flow rather than volume flow. Magnetic resonance is used to measure more phases than one and distinguish between them, facilitating the measurement of multiphase flow. But also well-established flow meters that have been in use for decades (such as Pitot tubes and bellows meters) still are and will remain to be in widespread use.

New developments do not only pertain to new measuring principles or types of flow meter. Flow meters are also increasingly equipped with on-board diagnostics that can let the engineer know that something might be going wrong soon or that they should arrange maintenance shortly. By diligent research much more has become known about the impact of the flow profile, swirl, and equipment geometry (particularly the need of leading and trailing undisturbed pipe lengths, bends, valves, contractions, etc.) on measurements. Furthermore, flow meters are being deployed in extreme conditions or with media that hadn't been thought of when those meters were invented.

Gas and liquid flows are being measured traceably over a very wide flow range, from the tiny (less than one millilitre per hour for drug delivery in intensive care units in hospitals), to the massive (tens of thousands of high-pressure cubic meters per hour in the case of fiscal metering of natural gas flows), and effectively everything in between. Moreover, increases in pressure and temperature range, and the addition of new media for which the flow should be measured, add to the complexity of flow measurement.

All flow meters have their vices and virtues in different conditions and with different media, and all have to be calibrated to give sensible results.

↑A large rotor gas meter which was used for calibrations in one of VSL's former laboratories in Dordrecht. (courtesy of VSL / photo by Cees Mastenbroek)

↑In one of VSL's so-called bell provers the Dutch cubic meter is born. This is the primary gas flow measurement standard in the Netherlands. Calibrations of flow meters using the left bell prover have a measurement uncertainty of below 0.1%. (courtesy of VSL / photo by Fons Alkemade)

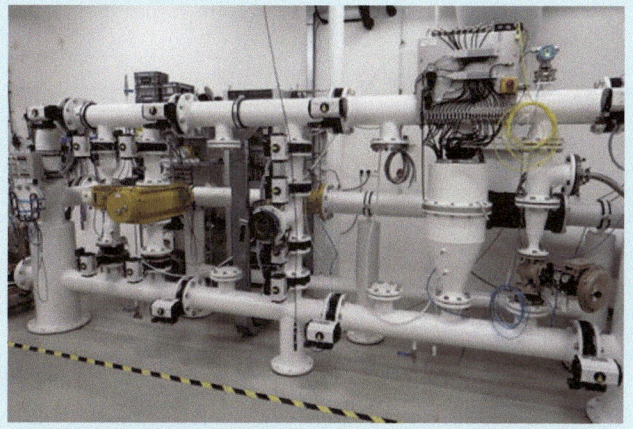

↑One of VSL's gas flow calibration facilities, showing various reference standards against which other flow meters can be calibrated. (courtesy of VSL / photo by Fons Alkemade)

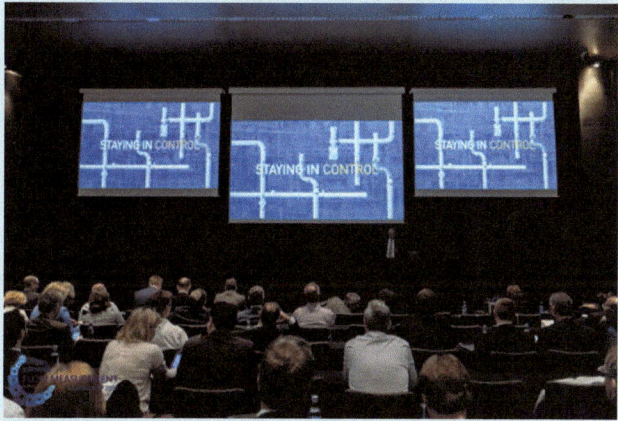

↑During the 6th European Flow Measurement Workshop, held in Barcelona in April 2018, nearly 200 delegates from 28 countries discussed how the newest developments in fluid flow measurement help them to stay in control of their processes. (courtesy of VSL / photo by Marcel Cloo)

In recent years, the energy transition has fuelled various new developments in flow measurement. The transport and storage of Liquified Natural Gas (LNG) takes place at -160°C, where many flow meters simply don't work. Systems where natural gas users not only receive gas, but also can supply to the grid (for instance farmers who generate biogas) have their own measuring challenges, if only because of the differences in composition of the various (even if small) flows involved. Hydrogen, seen as a key future energy carrier is such a small molecule that it is notoriously hard to measure at all. This was also shown clearly during the 6th European Flow Measurement Workshop which was held April 2018 in Barcelona, co-hosted by VSL, and where many exciting new developments in equipment and applications were shared. Indeed, everything flows: the field of flow measurement is itself in constant motion.

One question you might have: how does an NMI know that the values of the national standards are correct? What happens is that we compare our highest measurement standards to those of other NMIs in so-called round-robin tests, which are regularly held by groups of NMIs. These measurements are then compared, and the uncertainties of the measurements at the different NMIs are evaluated. Agreements are then made on how the various national standards compare to each other, and on the related uncertainties. The result of this is that calibration performed in the Netherlands should give the same results as those of a calibration in any of the other countries. By collaborating at the highest possible measurement level, countries make a joint best effort to make sure that the most accurate measurements are available everywhere and always, for everyone.

↑The Low-Speed Wind tunnel Laboratory around 1960. To the right was the main building of the Department of Mechanical Engineering. Behind the Low-Speed building the new Laboratory for Aerodynamics and Hydrodynamics would be finished in 1962. (courtesy of TU Delft/ photo by Fotografishe Dienst via Beeldarchief / CC BY)

↑Burgers' laboratory in Delft around 1930, with the largest, open-end wind tunnel on the left. (courtesy of Burgers Archives / Delft University of Technology)

→One of the two main halls of the 'new' Laboratory for Aerodynamics and Hydrodynamics, around 1990. At that time lasers had already become a familiar part of the equipment, and so were large sheets of plastic. Since 1962 the fluid mechanics engineers had plenty of space here, though the wind tunnels have never been very large. In the cellar most of the experiments involving water were done. The initial name of the Laboratory was still used when the Fluid Mechanics section had to leave this building in 2000 but officially it had already been abolished (Leen van Wijngaarden already called it an 'old-fashioned' name when he came to Twente in 1966). The building was completely demolished in 2017. (courtesy of Burgers Archives / TU Delft / photo by Fons Alkemade)

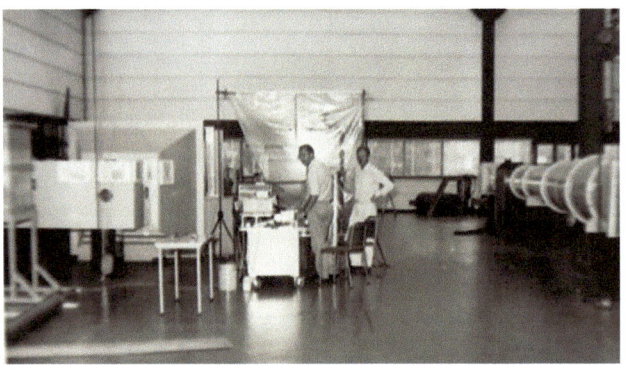

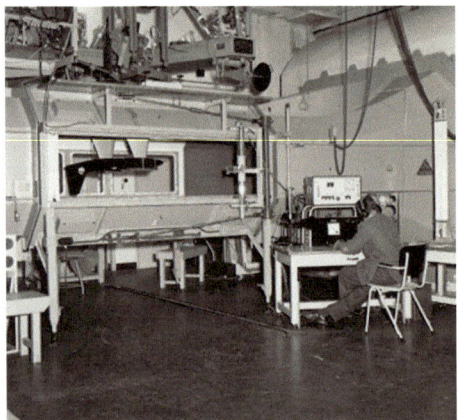

↑Measurements of an airplane model in the Low-Speed Wind tunnel Laboratory around 1958. Notice the headphones on the head of the researcher. (courtesy ofTU Delft / photo by Fotografishe Dienst via Beeldarchief / CC BY)

→The most recently opened facility at the LSL is this partly 3D printed vertical wind tunnel for aero-acoustic research. With wind speeds of up to 60 m/s, its height of 3.6 m and diameter of 3.0 m it is the largest of such facilities in the world. In this tunnel the sound production of e.g., the propellers of wind turbines can be investigated. For the designers of the tunnel, it was quite challenging to get the tunnel itself silent. Microphones are important measuring devices here.

## DELFT

As we have seen in § 2.3.3 the very first wind tunnel was in use around 1913 in Delft but it didn't survive the First World War. The first wind tunnel in Burgers' laboratory was ready for use in 1922. In 1927–1928 a second tunnel was put into operation. Around 1949 the Laboratory for Aerodynamics and Hydrodynamics had four wind tunnels. The largest of these couldn't be used for some time since its 20 Kw transformer had been confiscated by the Germans during the Second World War.

The first wind tunnel on the new 'campus' of the TU Delft, in the Wippolder, was not in the new Laboratory for Aerodynamics and Hydrodynamics of the Department of Mechanical Engineering. It was in the Lage Snelheids Windtunnel Laboratorium (Low-Speed Wind Tunnel Laboratory, LSL) of the Department of Aeronautical Engineering. This laboratory was also the very first building on the campus and when it was put into operation in 1953 one could still find grazing cows and small farms around it.

This low-turbulence wind tunnel, much larger than those in Burgers' laboratory, had been designed in cooperation with the NLR. It contained a quite revolutionary concept: the measuring section could easily be changed to another one. It was thus possible to have students doing their practical work (on a model of the Fokker T-5, for example) in the morning and to have researchers doing other research in the afternoon (or night). In 1977 the building was modified and extended. Among the new facilities was a unique vertical wind tunnel, which was named after professor Eise Dobbinga who had become Lecturer in Aerodynamics in 1954. Later a special boundary layer wind tunnel was completed.

Around 1960 research was done here on boundary layer suction; one of the researchers was Henk Tennekes who got his PhD in 1964 under the guidance of Steketee (see § 4.1). He then went to the USA and became known internationally after publication of the textbook *A first course in turbulence* which he wrote together with John Lumley and which was published for the first time in 1972. Later, Tennekes would become research director of the KNMI.

The LSL also became well known for its study on the prediction of the transition of two-dimensional flows around wings. In the course of time it also gained much knowledge on wing profiles for low speeds, and on those for sailplanes.

In 1959 a second wind tunnel was acquired by the still-young department. A supersonic tunnel (measuring section 15 x 15 cm) was installed in the cellar of the Laboratory. Initially it could only run for 90 seconds per day! From 1964 it was partly used for research on delta wings, a project related to the development of the Concorde supersonic airliner.

In 1967 the Department of Aerospace Engineering got its new main building in the southern part of the campus, where there was plenty of space to expand (see also § 4.1.1). Two years later a separate laboratory building, the HSL, with two high-speed wind tunnels was inaugurated. Today this Aerodynamics, Power and Propulsion Laboratory (containing both the LSL of 1953 and the HSL of 1969) has a number of high-speed and low-speed wind tunnels for aerodynamic measurements, including a low-turbulence wind tunnel, a large open jet facility and a 0.27 m diameter hypersonic facility. The open jet facility is a unique wind tunnel for testing model rotors of wind turbines. The hypersonic wind tunnel is one of the few in Europe based on the so-called Ludwieg-tube principle which makes it possible to reach Mach numbers between 6 and 11.

## RSL / NLL / NLR

Th second wind tunnel in the Netherlands was the first tunnel for the RSL. In 1918 its construction was started and on 5 April 1919 it was officially opened in a building on the Navy Wharf in Amsterdam (see § 3.2.1). Not much is known about the first measurements which were done in this tunnel of the 'Eiffel type'. The results were somewhat disappointing: the flow appeared to be not uniform and the maximum air velocity appeared to be 25 m/s and not the expected 33 m/s. Later in April 1919 there were already plans to insert a closed measuring section in the tunnel. This would be the first of a whole series of modifications.

For years this would be the only wind tunnel for the RSL. Only in 1938 was it decided to build two new tunnels, this time of the 'Göttingen type', a tunnel with a closed return circuit. In the biggest tunnel maximum speeds of 80 m/s could be reached. The new tunnels were installed in the new NLL building on the outskirts of Amsterdam and were almost finished when in May 1940 the Germans occupied the Netherlands. Ironically, the tunnels went into operation under the supervision of a 'Beauftragte', who reported on the developments at the NLL to the well-known aerodynamicist Betz, director of the Aerodynamische Versuchsanstalt (AVA) in Göttingen. In a discussion with Betz in July 1940 it was agreed that the NLL would not contribute directly to the war effort but could continue basic research activities in consultation with or even under contract from AVA.

During the War studies were done to transform one of the new tunnels into a facility for the study of high, but still subsonic, speeds. This appeared to be technically unfeasible and it was not until 1948 that new ambitious plans were announced: a huge High-Speed Tunnel (which became known as the HST) for speeds up to Mach 0.95; a low-speed low-turbulence tunnel for aircraft development; and a small compressor driven supersonic facility (the SST).

As we have seen in § 3.2.1 the construction of the HST could only be started in 1955. But the first supersonic tunnel, in fact the first in the Netherlands, was already operational in 1948. It had been designed by the German engineer Erdmann (see § 4.1.1), it had a test section of 3 x 3 cm, and could reach speeds of Mach 4.

After the completion of the HST around 1960, the NLL and its wind tunnels got an excellent reputation throughout the world. International cooperation developed, e.g., via the

  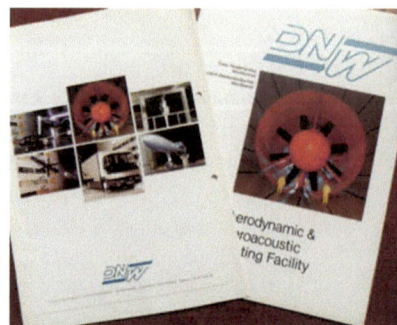

↑↗The Eiffel type wind tunnel of the RSL under construction at the Werkspoor company in 1918 and the official photo of the same tunnel made at the opening in 1919. Soon after the opening a range of modifications were made since the performance of the tunnel was somewhat disappointing. The entrance of this tunnel was changed to a so-called bell-mouth shape. (courtesy of Stichting Behoud Erfgoed NLR)

↑Selling the DNW services: brochures from about 1990.

→The shock tube at the NLL location in the Noordoostpolder in the early 1960s. Some years later the TH Eindhoven would also have a shock tube for research. (courtesy of Stichting Behoud Erfgoed NLR)

AGARD, the Advisory Group for Aeronautical Research and Development (related to NATO). Van der Maas appeared to be an excellent chairman of the NLL and accomplished much. In the Noordoostpolder, the new land which had been part of the Zuiderzee, a new location of the NLL was founded. Here the noisy experiments with the ramjet engines of the Dutch Kolibrie helicopter could be performed, as well as those with rocket propulsion. To study the phenomena which are involved in flows at high temperatures (where dynamic effects have very small time-scales) a shock tube was built there in the early 1960s.

In the 1960s the NLL also made plans for a new low-speed wind tunnel with a much larger test section than in the existing tunnels. For years nothing came of these plans and in the 1970s doubts were raised about the workload and hence the economical operation of such a wind tunnel. This led to

a cooperation between West Germany and the Netherlands and in 1976 a foundation was established, the Duits-Nederlandse Windtunnel (DNW). It was only in 1979 that the new facilities in the Noordoostpolder were ready for testing and calibration, and in 1980 the new impressive wind tunnel with test section of 9.5 x 9.5 m and two smaller ones were put into operation. Since then they have been used for the investigation of airplanes (sometimes full-scale) but also of cars, trucks, and other vehicles, for customers from Europe but also from countries like Brazil and China. The DNW tunnels have also been used for aero-acoustic investigations.

### EINDHOVEN EN TWENTE
Compared to Delft, the universities of Eindhoven and Twente never had many wind tunnels. Neither had aeronautical departments and in the late 1980s it was decided that Delft

←↑The HST of the NLL was – and is - impressive, both from the outside (see § 3.2.1) and from the inside. Its operation demanded a large increase of the output of the power plant of the institute. The operators and researchers of the tunnel were located in a special room. By 1960 one the means to get control of the operation of the tunnel was the use of closed-circuit television. (courtesy of Stichting Behoud Erfgoed NLR)

↑The first supersonic wind tunnel in the Netherlands around 1948, with Erdmann in the centre. In later years, an alternative test section was added as a pilot facility for the larger NLR wind tunnels. In 1967, the installation was donated to the University in Delft where Erdmann became professor. (courtesy of Stichting Behoud Erfgoed NLR)

would be the main location for turbulence research. In Eindhoven, from the late 1960s research was done with a shock tube by Rini van Dongen and others. This would become the start of a long tradition. In later years expansion shock tubes were used. Their main advantage is that moving parts are avoided and the fastest possible expansion of a gas can be realized. It can take pressures of up to 100 bar, which was chosen to make it interesting for the natural gas industry. In this tube nucleation was studied.

One wind tunnel in Eindhoven became well-known through newspaper articles and broadcasting in 2017. It is the huge Atmospheric Boundary Layer Wind Tunnel which allows the study of flows in a built-up environment on a model scale, but also full-scale studies, e.g., of racing cyclists.

In Twente a so-called freon tunnel was in operation in the early 1970s. Wind tunnels in which freon is circulating, instead of air, were already known in the 1950s. Thanks to the fact that freon allows a much lower speed of sound, this kind of tunnel can reach transonic flows more easily and this saves energy. Furthermore, much higher Reynolds numbers can be reached than in similar air-based wind tunnels. Freon tunnels have also been used for research on rotating machinery and flutter.

The only wind tunnel in operation today at the UT is mainly used for aero-acoustic research. It is known as the 'silent' wind-tunnel because its walls can absorb practically all the noise produced by the air flow, which can reach speeds of up to 240 km/h. This tunnel is suited for measuring simultaneously the aerodynamic behaviour and the noise production of applications related to aviation, drones, green energy production, and home appliances. Since this tunnel got a 'makeover' in 2017, it has its own Facebook page.

↑The freon tunnel of the TH Twente in 1971. Eindhoven also had a freon tunnel at that time. (courtesy of Beeldbank UT)

↑In the garden of the office of the Dienst der Zuiderzeewerken in The Hague a rather primitive wooden flume was erected in 1919 to study the wave run ('golf-oploop' in Dutch) on dike models. The man with the hat in the front is studying the wave generator. (courtesy for use of reproduction of Stichting Historie der Techniek, Eindhoven)

↑The official opening of the Atmospheric Boundary Layer Wind Tunnel in Eindhoven in December 2017 was a huge event. (courtesy of TU Eindhoven archives / photo by Bart van Overbeeke)

↑The open jet test section of the wind tunnel in Twente is located inside of a large anechoic chamber which has dimensions 6 × 6 × 4 m. Here research is done on the vortex generated by the flapping wing of a robotic bird. (courtesy of Kees Venner / University of Twente)

↑The opening of the renewed Deltagoot of Deltares in Delft, 2015. The high officials, seen on the right of the flume, were surprised by a huge tsunami-like wave and got wet. (courtesy of Deltares)

## •6.2.3 FLUMES AND BASINS

The first flume ('goot' in Dutch) for research purposes was probably that which was erected in the garden of the Dienst der Zuiderzeewerken (see § 3.4.1). It was 15 m long; the garden didn't allow a longer flume. Waves were generated by means of a simple hand-driven generator.

### WL / DELFT HYDRAULICS / DELTARES

In 1935 the WL had built a wind flume of 50 m length (later 60 m) in which waves were excited by air flows. Up till then only 'regular waves' had been generated in the traditional way by means of a 'board'. With this new unique facility (for years no similar one existed in the world) not only more 'natural' waves could be generated but also the energy transfer from air to water could be studied. Further, researchers could investigate the deformation of waves due to hydraulic structures (dams etc.) and the pressure caused by the waves on these structures. Thijsse was one of the main investigators in this field.

For many years the WL has had one major topic in its research: the Delta area. Even before the Storm Surge

↑↗Two examples of research done at De Voorst in the 1980s. As in Delft, De Voorst had a wave flume. The photo shows research in which the effect of waves on dunes is studied. The other photo shows a model of the Amsterdam-Rhine Canal and a model of a towboat with barges. The researchers wanted to know how this floating combination would find its own 'drift angle' while moving along the canal. (courtesy of Rijkswaterstaat / www.beeldbank.rws.nl )

↑One of the early models built in De Voorst, seen here in 1957: the situation near Arnhem where the river IJssel branches away from the river Rhine. (courtesy of Deltares)

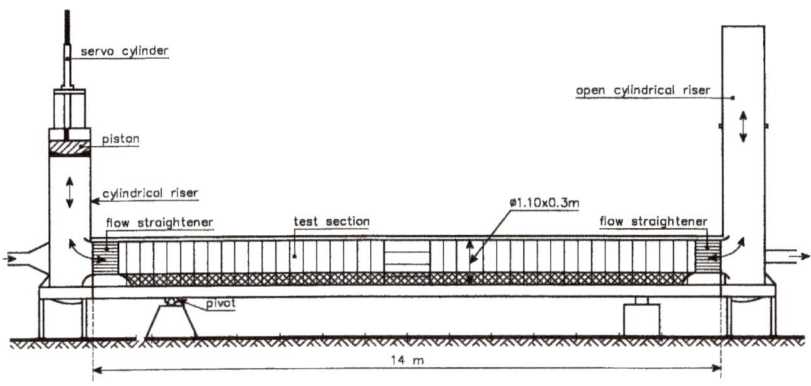

←This drawing from Jebbe van der Werf's PhD thesis *Sand transport over rippled beds in oscillatory flow* (2006) shows one of the remarkable flumes at Deltares. (courtesy of J. van der Werf)

Disaster of 1953 the WL had built and tested a rather large model of the northern Delta area, on the terrain next to its premises in Delft (see § 3.2.2). The purpose of the research was to discover the behaviour of tidal movements under different circumstances. Five minutes of flow in the real world could be simulated in about one second in the model and to investigate the happenings during a storm of four or five days it was enough to study the model for just twenty minutes. From 1951 onwards parts of the Delta area, e.g., the Haringvliet, were modelled in the annexe (dependance) of the WL which had been opened in 1951 in a corner of the

Noordoostpolder where there was still plenty of empty land. This second location was called De Voorst but became better known as the Waterloopbos. When it was closed down in 1995, it contained 35 large scale models of sea arms, harbours (that of Lagos, for example), and rivers. With these models the influence of hydraulic constructions like dams and locks could be investigated. From 1957 De Voorst had a unique 'wind- en stroomgoot' (wind and flow flume) in which it was possible to study the transport of granular material on the bed of a canal.
From the late 1960s more and more of the facilities in De

119

Voorst became covered by means of halls; the measuring equipment suffered quite heavily from wind and rain. After the closure in 1995, several of the models were left to nature and the 'ruins' can still be found in the Waterloopbos, which is now a protected monument.

On the new location of the WL in Delft (see § 4.2.2) one of the most remarkable flumes was the Delta flume (Delt-agoot). It was opened in 1969 and from the start large, full-scale waves could be generated and studied. Full-scale waves are a necessity in research since small waves cannot be compared to actual waves occurring in seas and rivers. The flume was rebuilt and deepened some years ago and reopened in 2015. Today, with a length of 300 m and a depth of 10 m, it is the largest wave flume in the world. Its wave generator can make waves of any required spectrum and is programmed to suppress waves which are reflected by the tested structure.

### DELFT

Some forty years after the start of the WL, the TH Delft opened its own laboratory for hydraulic research, the Stevin III Laboratorium or Laboratorium voor Vloeistofmechanica. In 1969 the huge concrete building (almost 100 m long and 18 m wide) was put into operation as one of the three laboratories attached to main building of the Department of Civil Engineering. In the early years it contained a flume of 30 m length and 0.60 m depth in which waves could be generated. One of the types of research which was undertaken in this flume was the behaviour of waves on slopes.

In the 1990s the TU Delft managers started to think about reducing the research area of the hydraulic researchers, and even of cancelling the whole laboratory, as square meters were expensive. After long discussions the reduction took place around 2010. The wave tank of 20 x 16 m had to be broken down. In 2014 a 'memorandum of understanding' with Deltares was signed in which it was agreed to exchange knowledge and researchers (including students).

Today the Waterlab, as it is now known, still contains several straight flumes, two long flumes (about 40 m) with a wave generator for the study of sediment transport, and a rotating annular flume. In a flume of 5 m width riverbeds can be studied. One of the subjects that can be measured there is the rheological properties of mud.

### WAGENINGEN

At the University of Wageningen (or Landbouwhogeschool at that time) the first flume was put into operation in 1950. It was also meant for practical work by students (up to 1956 all students in Wageningen had to follow a course in hydraulics during their first year).

The first primitive and very small laboratory for hydraulics in Wageningen was opened in 1959 but by 1963 it had been seriously extended. Research was done on spillways and

on models of groundwater flows. In 1970 the Laboratory for Hydraulics and Drain Hydrology got a new facility. It was a sloping flume of 15 m length of which the angle with the horizontal could be changed. It was initially intended for measuring the shear stresses on the walls of open ducts but soon a sprinkling installation was mounted above the flume for research on the drain of rainwater on paved surfaces. In 2010 a new laboratory was put into operation, containing several flumes, and it still has a rainfall simulator. The experimental research is mainly focused on morphological responses to channel flow and overland flow. The laboratory was named after its founder, Kraijenhoff van de Leur (who had been a student of Thijsse in Delft).

### UTRECHT

The Physical Geography Laboratory at the University of Utrecht has a soil and sediments lab, an experimental flume lab, and also the Metronome. To create reversing tidal flow and sediment transport on scale, professor Maarten Kleinhans built a 20 m long and 3 m wide flume tilting back and forth every 30 seconds at a small slope. This tilting-principle causes Waddensea-like basins and estuaries to form in a pilot setup with sand created from a small table with one leg sawed off and replaced by something like a metronome. Using this immense research instrument, Utrecht researchers simulate how river mouths are formed.

For some years Kleinhans and his team also studied possible flow patterns on the planet Mars. Experiments with water and sand showed that an eruption of groundwater can be held responsible for what is visible on the surface of Mars. Study of photos made on the planet and calculations led to the conclusion that water had been flowing not for thousands of years but for much shorter periods.

## • 6.2.4 TOWING TANKS

### DELFT

The very first facility which Burgers could use in his laboratory can be regarded as a towing tank. It was 8 m long and a small cart could be pulled along the tank on top of it. Objects like cylinders could be drawn through the water and a camera on the cart was used to take photos (see § 6.2.1 for some results of the experiments). The tank could also be used as a water tunnel in which water could be pumped around at a maximum speed of 12 m/s. Further, the tank could be used as a flume in which waves could be generated. This towing tank was demolished in 1934 to make room for another wind tunnel.

As we have seen in § 4.1.1, the first towing tank (or model basin) at the Department of Naval Architecture in Delft was built in 1937. It had a length of 37 m and a width of 2.7 m. Towing was done by an unmanned small carriage carried by

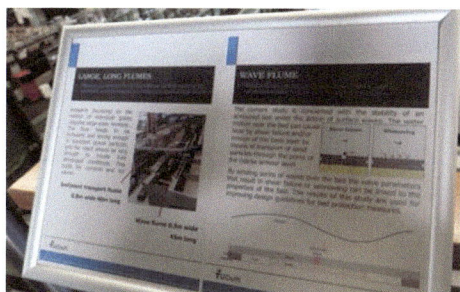

↑↗→ The Waterlab today. Visitors have a nice view of all research activities from the balustrade which encloses the large hall. On the balustrade several instructive signs tell them about the facilities below.

↑ The sloping flume in Wageningen around 1971, with the 'rain simulators' above it. (courtesy of Wageningen University & Research)

↑ The Metronome in Utrecht. (courtesy of Utrecht University / photo by Jarno Verhoef)

↑ The towing tank/water tunnel/wave flume in Burgers' laboratory in Delft around 1930. (courtesy of Burgers Archives / Delft University of Technology)

↑ The long towing tank of the TU Delft today, with tested models on the wall. Besides the long and the small towing tanks for research, a much smaller tank is in use today, mainly for the training of students. It has been named Bruno, after the towing tank pioneer Tideman (see § 2.3.1).

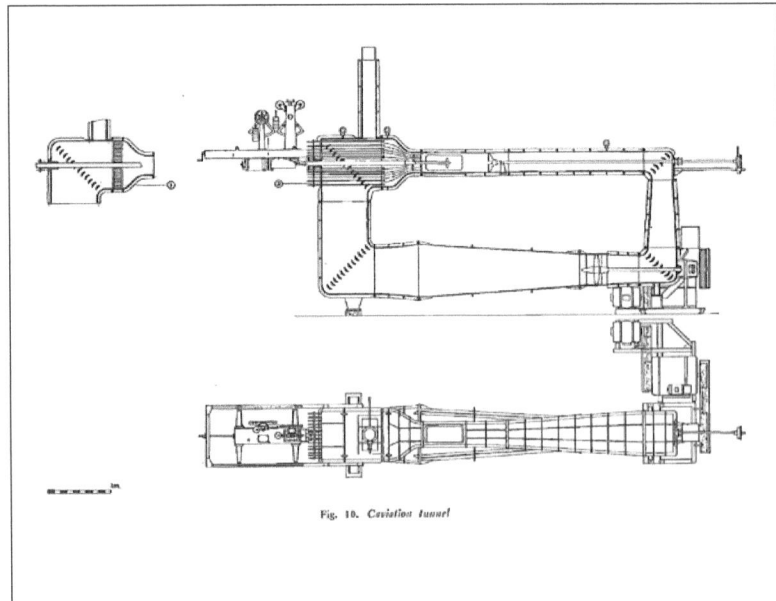

Fig. 10. Cavitation tunnel

←The cavitation facility in Delft as it was built around 1956. The so-called elbow number 1 was used when a propeller had to be investigated in a homogeneous velocity field. A specified velocity distribution over the screw disc analogous to that occurring behind a ship could be realized by using elbow 2. This non-homogeneous field was produced by a velocity regulator, containing 146 elements. By means of some type of check valve each of the elements could be (more or less) shut off, which affected the velocity of the water flowing through the elements. The tunnel had two cross sections, both of 300 x 300 mm. One was for measuring friction drag reduction by air lubrication, the other for measurements on cavitation vortices. (from: Gerritsma (1957) / with permission from IOS Press / publication is available at IOS Press through http://dx.doi. org/10.3233/ISP-1957-43001)

←For many years all activities in the Ship Hydromechanics Laboratory have been recorded on hundreds of photos. This is one of the impressions made in 1958, when the long towing tank had just been put into operation. (courtesy of Delft University of Technology / Cornel Thill)

↑The Wave and Current Basin of the NSP in the late 1960s. (courtesy of MARIN)

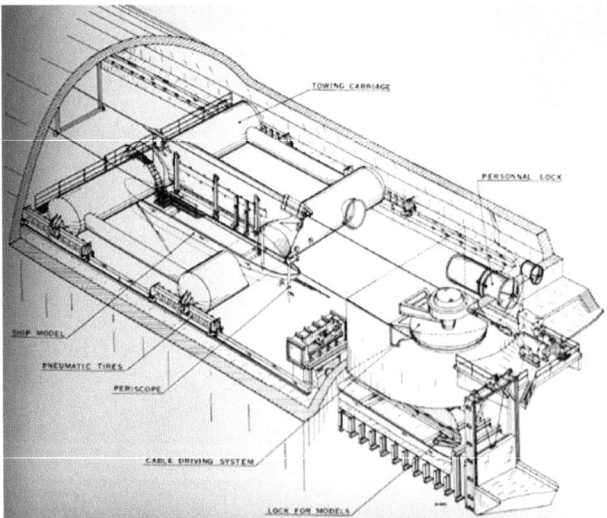

↑The Vacuum Tank of the NSP as it was originally built in 1972. (courtesy of MARIN)

↑The Twente Water Tunnel. (courtesy of Physics of Fluids group / University of Twente)

↑The cavitation/water tunnel of the TU Delft as it is today. Here research is done on drag reduction by means of air lubrication along a flat plate. Measurements are performed with PIV.

a rail over the tank. This carriage carried a resistance dynamometer with which the resistance of a model (maximum length 1.5 m) could be measured. In 1953 a wave generator of the 'flap type' was installed with which waves with a length of 3 m and a height of 9 cm could be generated. Around 1956 a new tank was built, with a length of 97 m and a width of 4.3 m and an 'ordinary' water depth of 2.5 m. This tank was equipped with a wave generator of the pneumatic type. In 1957 the staff of the tank prepared for the construction of the first electronic dynamometer, for momentarily measuring torque and thrust.

Since 1957 the Ship Hydromechanics Laboratory encompasses three test facilities: one large tank (142 m long, model speeds up to 8.0 m/s), one small towing tank (75 m long), and a cavitation tunnel (see § 6.2.5). All have been exploited in the context of hydromechanical research both in the maritime and offshore domain and are extensively used for education, scientific research, and cooperation with industry, both nationally and abroad. Beside these, there are two very small towing tanks, mainly used by students.

### NSP / NSMB / MARIN

As we have seen in § 4.2.3, in the 1950s and 1960s the NSP was expanding. The great demand for specialist research required the building of more sophisticated test facilities. The first one was the Seakeeping Basin (1956) for research into ship behaviour in waves. Next came the Shallow Water Basin (1958), for tug boats, inland shipping, and merchant shipping in shallow water. In 1965 a High-Speed Basin and a Wave and Current Basin were put in operation. In the Wave and Current Basin (today known as the Offshore Basin) wind, current, and waves were simulated for research into the behaviour of structures during complex operations at sea, such as oil and gas production and dredging operations. In 1972 the institute was further extended with a Vacuum Tank (not in Wageningen but in nearby Ede), which had dimensions 240 x 18 x 8 m and in which the air pressure can today be lowered to 2500 Pa. This facility was and still is used for research into 'wave impacts with air entrapment' and propeller cavitation. The need for low pressure comes from the rules imposed by model experiments: the smaller a ship model with its own rotating propeller, the smaller the air pressure should be.

In 2000 and 2001 the institute was (again) thoroughly updated. The Seakeeping Basin and Wave and Current Basin were replaced by a new Seakeeping and Manoeuvring Basin (170 x 40 m) and a new Offshore Basin. The first is one of the largest testing facilities in the world for this type of testing. It is designed for making arbitrary (high-speed) manoeuvres in realistic waves from arbitrary directions. Models of up to 8 m can be tested. The Vacuum Tank was completely renovated,

123

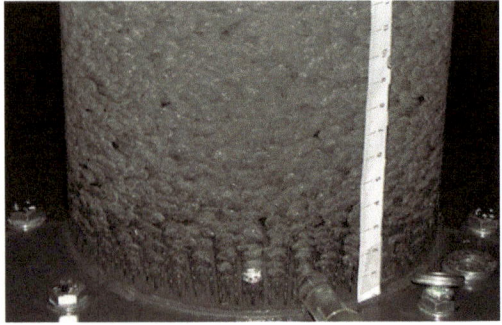

↑→The bubble column in Delft as it is today. It was the first in its kind with so many needles and such large diameter. Besides, the bubble flow remained uniform for a long time; only at a volume fraction of 55 percent the water started to circulate. (courtesy of TU Delft / Rob Mudde)

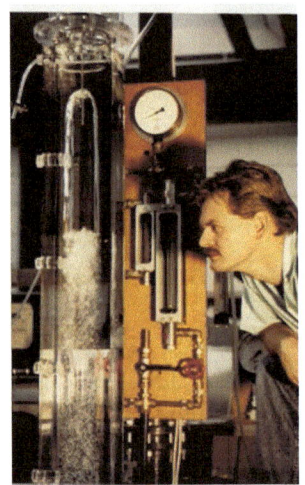

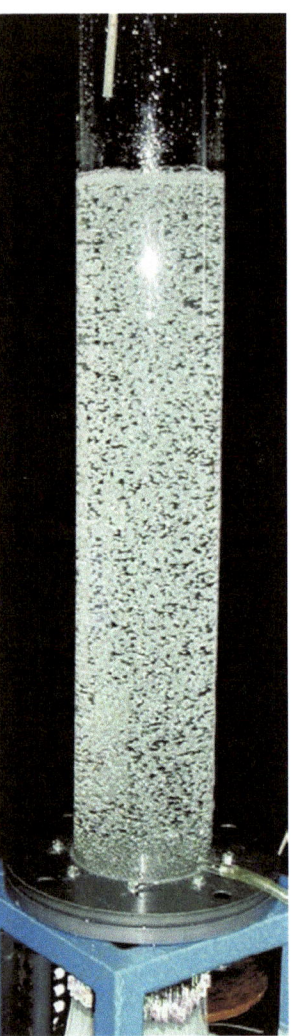

←The facility at the Laboratory for Aerodynamics and Hydrodynamics of the TU Delft in which Taylor bubbles could be observed, around 1994. (courtesy of TU Delft / René Delfos)

↑Schmid's experimental facility in Delft around 1929. (from: Schmid (1930))

getting flap-type wave makers on two sides of the tank; it is now called the Depressurized Towing Tank.

## • 6.2.5 WATER TUNNELS

As we have seen in § 6.2.4, the first experimental facility in Burgers' laboratory (1921) was a towing tank which could also be used as a water tunnel. There is however no evidence that it was ever used as such. It is known that in 1933–1934 this laboratory had an experimental setup with which the flow through water pipes and cavitation could be investigated. At the end of the 1930s Burgers was asked to advise on the design of the cavitation tunnel which the NSP was planning to build (see § 3.2.3).

This tunnel of the NSP was put into operation in the 1940s. It meant a breakthrough in the research on the cavitation phenomenon which occurred around ship propellers. Some years later, around 1956, a cavitation tunnel was put into operation at the new Shipbuilding Laboratory of the Department of Mechanical Engineering and Naval Architecture in Delft (see § 4.1.1). This laboratory also had a flow channel which was 45 m long, 2.8 m wide and had a water depth of 0.6 m. The water in this channel was pumped around with a velocity of up to 2.5 m/s but experiments in stagnant water were also possible. The flow velocities in this channel were measured with either a Pitot tube (which was not so accurate) or a small propeller.

A completely different kind of water tunnel, and much more recent, can be found at the University of Twente. The Twente Water Tunnel is an 8 m high facility in which strong turbulence can be created using a so-called active grid. Light particles like small bubbles and hollow spheres can be suspended in the turbulent flows. These particles may be made to rise with or against the flow and can be observed and followed in the measuring section for long-duration tracking.

## • 6.2.6 OTHER FACILITIES

Besides wind tunnels, flumes and basins, towing tanks and water tunnels, many other experimental facilities have been built during the last hundred years for the study of flows. Below a short and rather arbitrary selection of these.

↑←Two experimental facilities at the dredging laboratory in Delft. The pipeline has a total length of about 60 m and has been used to investigate the wear of its inner walls caused by slurry flows. The facility with the sloping bottom is used to do research, using acoustic devices, on the downhill flow of slurries

↑The mast near Cabauw has 9.4 m long 'protrusions' which are placed at distances of 20 m. With these it is possible to avoid disturbances caused by the tower during the measurements. (courtesy of KNMI / Wouter Knap)

## MULTI-PHASE FLOWS

The gas-lift, or air-lift, technique is a gravity-driven pumping process applied in the oil industry for many years. It enhances oil production by injecting gas (air) in the production pipe to decrease the hydrostatic weight of the oil column: the rising gas takes the oil with it, as it were.

At the end of the 1920s Burgers was consulted by a PhD student of one of his colleagues at the Department of Mechanical Engineering. Wilhelm Schmid, who would become professor in Eindhoven in 1957, was working on measurements on an air-lift which was built outside the building of the Department. It was a huge gantry of about 20 m height which held a tube with an inner diameter of about 5 cm. Water was taken from a source which was 36 m below ground level. In this tube Schmid was able to let single, rather large air bubbles rise and was able to measure the rising speed. His thesis (Schmid (1930)) can be considered as one of the first Dutch publications related to two-phase flows.

Some seventy years after Schmid's research, gas-lift was again studied in Delft. In 2004 a PhD thesis entitled B*ubble*

*size effect on the gas-lift technique* appeared, based on experiments with a vertical upward bubbly pipe flow, with a height of 18m and a diameter of 72mm. The flow velocity conditions investigated in the experiments were representative of practical gas-lift circumstances. The work was supervised by professors Gijs Ooms (see § 4.1.1) and René Oliemans (who had been working on multiphase flows at Shell's KSLA from 1982 and at the TU Delft from 1990). Some ten years earlier related research was done by René Delfos in Delft, also under the supervision of Oliemans. He used an experimental set-up in which he could realize a stationary so-called Taylor bubble. With this he could study the phenomenon of gas-liquid slug flow, a phenomenon well-known in the oil industry where efficient transport of oil-gas mixtures through pipelines is required.

Two-phase flows of many bubbles rising in a liquid are well-known from chemical reactors and were the subject of research at the Kramers Laboratory (see § 3.3.2 and 4.1.1). From about 2000 the construction of a so-called bubble column was built, containing a sparger (bubble generator)

125

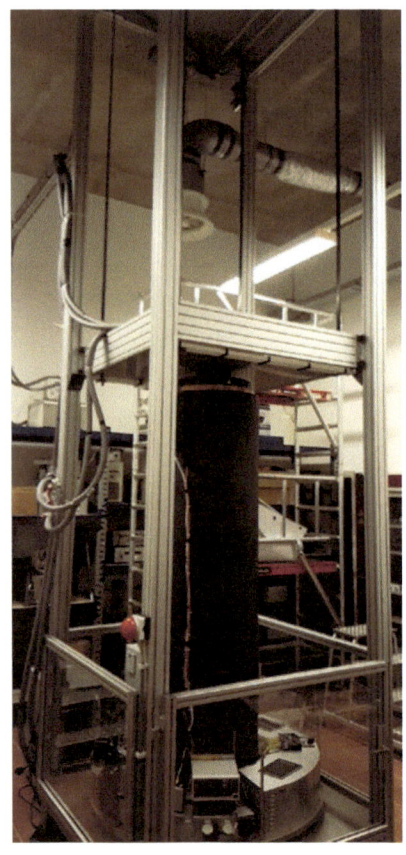

↑For his experimental research Hooghoudt built this huge facility in the late 1930s at the Rijkslandbouwproefstation in Groningen where all kinds of agricultural research were performed. (from: De Vries (1982); courtesy of Co de Vries)

←The TROCONVEX in Eindhoven was still growing in 2018. (courtesy of TU Eindhoven / Rudie Kunnen)

with 560 special needles from which bubbles with virtually all the same velocity and size could rise. This facility has been used to investigate the conditions under which large scale structures arise; these are usually undesirable in industrial reactors.

Quite different two-phase flows occur in the world of dredging. Dredging is a combination of fluid mechanics and soil mechanics (rock, mud) which makes modelling very challenging. Research into the flow phenomena related to dredging started in 1937 when a laboratory was opened in Haarlem (after the TH Delft had refused to make laboratory space available). It was used to do experiments on the suction of sand. It wasn't till the end of the 1960s that serious research on the dredging process was started in Delft. At that time the WL was already doing experiments in this field, mainly related to the Delta Works.

For several years the TU Delft has contained the Dredging Engineering Research Laboratory, one of the very few in its kind around the world. Several PhD students have done their research there, using the unique facilities in which slurries can flow. Some of the big dredging companies in the Netherlands have their own experimental facilities but in Delft the more fundamental research is done.

### GEOPHYSICS

During the 1960s the need for more data about the atmo-

sphere led to plans for a new facility. This led to several attempts by the KNMI to find existing masts and build new ones from which measurements could be performed. Since its construction in 1972 the mast near the village Cabauw (213 m high, 2 m diameter) has become a leading atmospheric observatory in the world. The first measurement program with this mast, in 1973, was related to air pollution. The observatory is one of the few that can characterize the atmosphere from the ground up to the top of the atmosphere, by combining in situ sensors on the ground, along a 213 m measurement tower, and ground-based remote sensing to reach higher altitudes for measuring wind, turbulence, aerosols, trace gases, and clouds and radiation.

Since 2002 the facility became known as CESAR: Cabauw Experimental Site for Atmospheric Research. The station has become a focal point of experimental atmospheric research in the Netherlands, and a core station in the global network of observatories. The observatory provides essential state-of-the-art data to understand atmospheric processes, validate satellite observations, and detect long-term trends. The freely available data are used to improve climate, weather, and air quality models. Today, three universities and five major research institutes collaborate in CESAR. Thus, the facility has a strong integrating effect on atmospheric and environmental science in the Netherlands.

Many geophysical and astronomical phenomena are driven by highly turbulent fluid dynamics. These dynamics are often

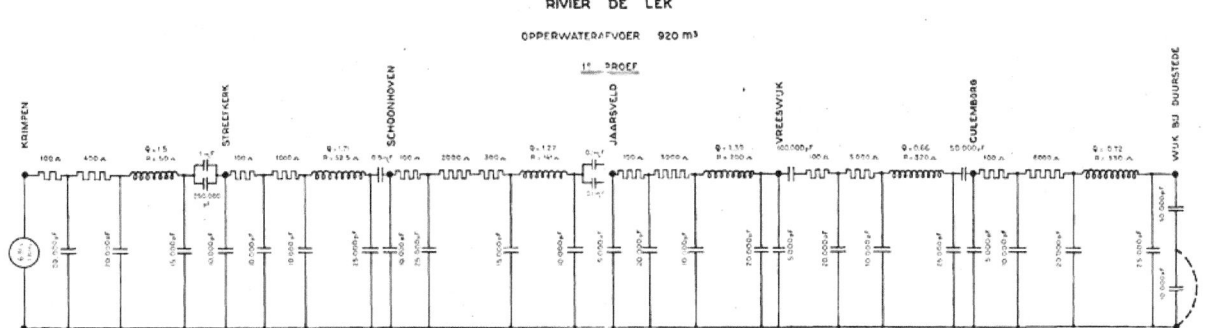

RIVIER DE LEK

OPPERWATERAFVOER 920 m³

1ᵉ PROEF

↑A river transformed into an electrical circuit. This was the kind of analogon that Van Veen proposed to RWS and finally led to one of the first analogue computers in The Netherlands. (Scheme from 'De voortplanting van het getij bepaald met behulp van de electrotechniek', a report prepared by H.J. Stroband for Van Veen in 1944/1945. (courtesy of Min. van Infratructuur en Waterstaat).

←Women played an important role in scientific calculating during the 1930s until the 1950s. At the start of the Mathematical Centre in Amsterdam a group of girls who had just left high school were hired and trained to become 'calculators'. They not only learned to use the mechanical and electro-mechanical calculators of those days but also to operate the first computers. This photo was taken around 1955 In Burgers' laboratory in Delft. It is not known whether this lady was especially hired to do scientific calculations. (courtesy of Burgers Archives / TU Delft)

driven by the buoyant rising and falling of fluids of different densities, known as convection (see § 5.1.7), and strongly affected by the rotation of the celestial body through Coriolis forces. One useful approach to understanding geophysical and astrophysical flows is to study a reduced problem known as rotating Rayleigh-Bénard convection in a laboratory setting.

While studies of rotating convection began over a century ago, only recently have major developments occurred toward understanding the problem in settings of exceptionally strong rotation or convection, such as geophysical systems. Modern laboratory and numerical studies have found that many novel behaviours emerge only under extreme conditions.

One of the facilities built to do research under these extreme conditions is the TROCONVEX (Turbulent Rotating Convection to the Extreme) at the TU Eindhoven. With a maximum tank height of 4 m, the rotating TROCONVEX can generate stronger convective and rotational forces than any other laboratory rotating convection device to date (2018).

### DRAINAGE FLOWS

As we have seen in § 6.2.1 groundwater flows and drainage have long had the attention of quite a number of Dutch scientists. Besides theoretical and numerical research, experiments also have a long history in the Netherlands. Scientific research on land drainage was stimulated by the wish to reclaim and develop the new Zuiderzee polders in a rational manner. This successful research programme began in the 1930s under Sijmen Hooghoudt (1901–1953), a chemist by training, at the Experimental Station and Soil Science Institute in Groningen. This led to a basic understanding of the drainage process, and the emergence of extensive basin-scale studies of groundwater discharge regimes, and the related drainage requirements for different hydrologic and topographical conditions, in connection with the improvement of land and water management. Hooghoudt also became known for a formula, named after him, which is still used in drainage studies.

## 6.3 NUMERICAL APPROACHES

The ability to calculate flows has in part replaced experiment and has become an essential part of research into the fundamentals of fluid flows. Furthermore, in the engineering design process, it allows for rapid evaluation of changes in design parameters. Many believe that the numerical approach only started when the digital computer was maturing in the 1960s but in fact this approach was already used before the Second World War when even the analogue computer didn't exist.

←The Electric Model of Waterways around 1960 and one of its operators. During the design of the machine, RWS came to realize that it would not fit in one of their offices and therefore they had to find a place somewhere else. It was finally installed in the building of the Freemasons in The Hague. (courtesy of beeldbank. rws.nl, Rijkswaterstaat)

## • 6.3.1  PIONEERS

One of the places where tough and long-term calculations were necessary in order to predict flows was Rijkswaterstaat (RWS). The work that had been done by the Lorentz Committee on the closing of the Zuiderzee (see § 3.4.1) had shown that the amount of calculation work was enormous. There was one engineer at RWS who was convinced that there had to be an alternative which would give results much more quickly: Johan van Veen (1893–1959). Van Veen has become one of the best-known engineers in The Netherlands. Soon after he entered RWS in 1929 he was put on the study of sand transport in the North Sea. Van Veen could make use of the measurements that were done in 1933 by the RWS measuring ship De Oceaan and was able to finish his thesis in 1936, becoming one the very first, and for decades one of the few, civil engineers to become a doctor in science.

At the end of the 1930s, Van Veen saw similarities between the formulas for electric currents and those of tidal currents. And since electrical formulas rule the change of current, voltage, etc., in electrical circuits, he suggested building models of waterways from wires, resistances, capacitors, etc. Van Veen showed that it worked (to a certain level of exactness) and that he could replicate the results which Lorentz and his Zuiderzee Committee had found.

But not everyone was convinced of the value of this method. RWS refused to use Van Veen's method to do calculations on the lower rivers. Instead, they used the so-called 'exact method' which had been developed by the mathematician

Jo Dronkers (1910–1973), the first to have been hired by RWS. Dronkers managed to refine the approach which had been used by Lorentz and his Zuiderzee Committee to calculate the influence of the construction of the Afsluitdijk (see § 3.4.1). Whereas Lorentz had used a 'linearized resistance' in his model, Dronkers used a quadratic one and was able to simulate tidal movements much more accurately. With this approach to tidal movements the Netherlands took a leading position in the world. In 1964 his influential book Tidal computations in rivers and coastal waters was published. Van Veen's 'calculations' with his electrical analogon circuits led him to the conclusion that the Netherlands should seriously fear the high water levels which would occur during heavy storm surges. The dikes in Zeeland province and other areas would not be able to withstand these levels. Others were not convinced that his figures were correct and were of the opinion that the situation would never get so bad. The infamous Flood of 31 January – 1 February 1953 would show that Van Veen's concerns had not been exaggerated...

## • 6.3.2  THE FIRST COMPUTERS

In the Burgers' Laboratory annual report for 1945–1946 we read how, slowly, the facilities and equipment were reinstalled and supplemented. There was a gift from the Department of Marine: high pressure compressors and reservoirs from demolished submarines which could be useful for the starting research on high-speed flows. Furthermore, some books could be bought. Also mentioned are the components which could be bought in England thanks to the Help Holland Council, a kind of relief fund from Great Britain. Burgers

The first computer of the NLL, the ZEBRA, was used from 1958 for data reduction of the High Speed Tunnel. (courtesy of Stichting Behoud Erfgoed NLR)

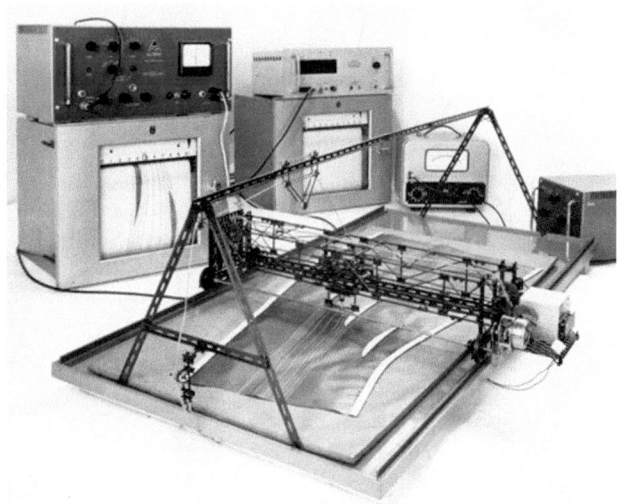

↑Around 1963 this analogue machine was used in the Physical Laboratory of the Stork company for research on the flow around the fans of pumps and similar machines. (courtesy of Historisch Centrum Overijssel / Fotocollectie Stork)

managed to acquire capacitors and transistor lamps for his optical equipment and also arranged a deal with the famous Meccano factory: it would produce some extra boxes of Meccano parts (against 'pre-war prices') for the 'differential analyser according to Hartree' which the Delft group wanted to build.

This, at first sight rather remarkable, purchase shows something of the situation around 1945 with regard to the practice of scientific calculations. The first primitive computers had just been built but these were still far out of reach for the academic world, for institutes, and industry. They were also still of little value since their performance was still very poor. It is not known whether the Hartree-machine was ever built (and used) in Delft. Maybe the building of such a machine was harder than the Dutch scientists had thought. Or perhaps they decided to abandon this project since they were told that much better electronic calculating machines would become available in the near future.

One of the Delft engineers who was convinced of this, was Aad van Wijngaarden (1916–1987). He had been a student of naval architecture in Delft and became involved in research done at Burgers' laboratory during the War. In an interview Van Wijngaarden later recalled: "It wasn't nice. It didn't give the level of insight that I wanted. I tried hard to do certain calculations; these were calculations for which you wouldn't get anyone today without offering him a computer. It was wartime. Day in day out I was behind my Marchant electric calculator at home; that saved me the time that was needed to go to the lab and back. Rising early in the morning, calculating the whole day and back to bed late in the evening.

That is how it went, week in week out. … It was about a problem from fluid mechanics: boundary layer equations. These were non-linear third order partial differential equations. It was huge task, you could only come further by just going on and on. … After some weeks I had finished one of the equations." [translation by FA]

Finally, Van Wijngaarden decided to leave the Burgers group and wrote his PhD thesis under the guidance of Biezeno, the professor of solid mechanics. After the War Van Wijngaarden became aware of developments in computing when in 1945 the Help Holland Council supported his visit to England. In 1946 Van Wijngaarden became the director of the 'calculation department' of the recently founded Mathematical Centre, the precursor of the present CWI. In 1947 he and some other young men started to build the first Dutch computer, the ARRA. The second version of this machine would become the first computer used for fluid mechanics: the Fokker aircraft company bought the FERTA (Fokker's Eerste Rekenmachine Type ARRA) in 1955, mainly for calculations in aerodynamics.

Despite the doubts about Van Veen's ideas, RWS decided to build an electrical analogon machine for calculations of the (tidal) flows in the lower rivers of the Netherlands. The machine was called Electrisch Model van Waterlopen (Electric Model of Waterways) and it was completed in 1954. It never worked very satisfactorily and in 1955 RWS had already decided to build a new analogue computer.

The new machine became known as the DELTAR: the Delta-tij-analogon-rekenmachine (Delta-tide-analogon-calculator). The Deltar has become known as one of the first

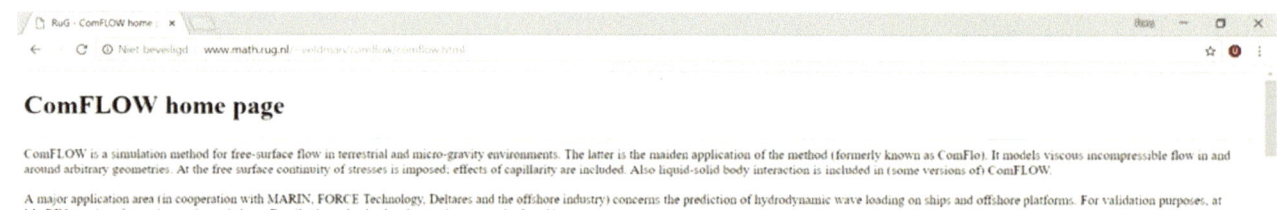

### ComFLOW home page

ComFLOW is a simulation method for free-surface flow in terrestrial and micro-gravity environments. The latter is the maiden application of the method (formerly known as ComFlo). It models viscous incompressible flow in and around arbitrary geometries. At the free surface continuity of stresses is imposed; effects of capillarity are included. Also liquid-solid body interaction is included in (some versions of) ComFLOW.

A major application area (in cooperation with MARIN, FORCE Technology, Deltares and the offshore industry) concerns the prediction of hydrodynamic wave loading on ships and offshore platforms. For validation purposes, at MARIN a series of experiments is carried out. Details about the dambreak experiments can be found here.

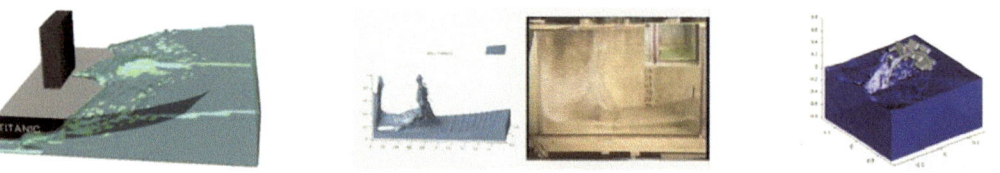

Another application area (in cooperation with NLR) concerns spacecraft dynamics as influenced by onboard liquids. Validation experiments have been carried out in February 2005 with the Sloshsat FLEVO satellite.

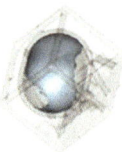

Finally, a ComFlo version has been developed to model hemodynamics in elastic arteries.

↑As academic groups developed software which was interesting for many users, they started to offer their services. The origins of ComFLOW of the CFD group of the University of Groningen can be found in their activities during the late 1990s. Other groups are also offering their software on the internet, e.g., the Multiphase and Reactive Flows group of the TU Eindhoven. (courtesy of University of Groningen / Arthur Veldman)

successful analogue computers in the Netherlands. It certainly proved to be helpful in the prediction of the behaviour of the Delta area but soon RWS had to conclude that doing calculations on digital computers, based on the mathematical models proposed by Dronkers, was a better way to go forward. Despite this, the DELTAR was used until 1982.

In 1958 the first Dutch transistorized computer reached the market and from then new developments followed each other quickly. That same year the NLR would buy its first computer (see also § 4.2.4) and one year later the first Dutch university would do the same. At the start of the 1960s many Dutch scientist and engineers became convinced that the computer would soon be a wonderful and indispensable tool for their work.

"The use of mathematics and computers has already been mentioned as an indispensable aid. The rapid developments in this field give reason to investigate whether some hydraulic problems solved hitherto only by means of models can be solved entirely or partially by means of computer techniques. Unsteady flow in networks of canals and flow in two horizontal dimensions under the influence of the topography are already, in theory, fit for treatment in computers. It is only a matter of the training of the programmers and the availability of computers." This was written in... 1963 by Harold Schoemaker (1913–2011), Thijsse's successor at the WL in Delft (and professor for irrigation studies at the TH Delft from 1967). It shows the enormous optimism about the

developments of computational fluid dynamics (CFD) during the 1960s. But as for the WL (and other research centres with experimental facilities) it would take some decades before the first model basins were dismantled as faith in the reliability of computer models grew to a high enough level.

### • 6.3.3 CFD

During the 1960s institutes like the WL and NLR started to develop their software for numerical simulations of flow problems. At the end of the decade, activities in what would become known as CFD (Computational Fluid Dynamics) also reached the academic world.

At the University of Leiden there was one location where a somewhat special brand of fluid mechanics was studied. It was at the Leiden Observatory in the late 1960s that the student of astrophysics Bram van Leer got interested in CFD for the sake of solving cosmic flow problems, especially shock waves. From the 1960s flows in shock tubes could be solved by a method disclosed by the Russian scientist Godunov, but this method showed wiggles: instabilities in the results which are physically wrong. Van Leer came up with an improved model which was more accurate, more efficient, and showed no wiggles. His method also became much used in aerospace aerodynamics. In the 1980s Van Leer emigrated to the USA where in 2010 he received the AIAA Fluid Dynamics award for his lifetime achievements.

In the 1970s many groups around the country were working on CFD and in 1974 some of these started the Kontaktgroep

Numerieke Stromingsleer (Contact Group Numerical Fluid Mechanics). Its purpose was "to provide an opportunity for researchers in numerical fluid mechanics to meet regularly and to inform each other about their research in an informal atmosphere."

It was Pieter Zandbergen (see § 4.1.3) who played an important role in the rise of CFD in the Netherlands. Around 1985 he took the initiative towards a joint effort in CFD which emerged from his NLR connection and led to the the ISNaS project (Information System Navier-Stokes). Subsidised by the Dutch Ministry of Education and Sciences and the Ministry of Transport, this project established a strong collaboration between two universities (Twente and Delft) and two laboratories (NLR and WL), with Zandbergen in the chair of the project control board (1987–1993). Later the University of Groningen, ECN, and MARIN, also joined the project. The ISNaS-project aimed at providing tools for computer-aided design and engineering in aerodynamics and hydrodynamics by developing an information system for the simulation of complex flows based on the Navier-Stokes equations. Major components of the project were the development of a so-called method-shell and of accurate as well as robust solvers for both compressible and incompressible flows. For the incompressible case, guided by typical applications in the field of river and coastal hydrodynamics, a solution procedure was developed which was capable of handling complicated geometries, including free surface effects, in particular for high-Reynolds number flow regimes. Within the ISNaS project a strong need for adequate turbulence models became apparent and consequently, Zandbergen and his group started investigations on the issue of turbulence and large-scale computations in 1990. In line with the subject of large-scale computing in 1993 Zandbergen became chairman of the national initiative on High-Performance Computing and Networking.

Together with the improvements which were reached in the development of software, big steps were also made in hardware. Besides 'parallel computing' the word 'supercomputer' became one of the buzzwords in the 1990s. In 1996, for example, the TU Delft proudly announced that it had installed the most powerful computer in the Netherlands, a Cray T3E for parallel computing. Among the main users of this supercomputer were the research groups in fluid mechanics, especially those working on turbulent flows. This Delft record, by the way, didn't last very long as soon an even more powerful computer was installed elsewhere. Some research groups started to develop their own hardware, special one-off computers. The group of Hoogendoorn in Delft, for example, built a processor to directly solve the convection-diffusion algorithm for the finite control volume method as well as the transport equations. It was called the Delft Navier-Stokes Processor (DNSP) and it appeared that the DNSP could execute certain calculations with a speed

comparable to those of the supercomputers used by others. The finite-volume flow solvers developed within the ISNaS project had an appreciable impact on the research of the participating university groups. Piet Wesseling at the Department of Applied Mathematics in Delft (originally an aeronautical engineer) became an expert in so-called multigrid methods and wrote a well-known textbook on the topic in 1991. Arthur Veldman (whose career started at the NLR) and his team in Groningen succeeded in the simulation of 2D turbulent flows by using DNS, but around 1995 this still required clever solutions for the spatial discretization problems. Numerical methods were also developed at the Centre for Mathematics and Computer Science in Amsterdam (CWI). Some research groups started to use 'commercial software', of which Fluent was and is a well-known example. Others mistrusted these packages or discovered that they lacked the 'precision' and applicability which was needed for their flow problems; they developed their own software. Thus, RWS came up with its WAQUA package for the simulation of flows in waterways and for the transport of sediments. For the ECN team working on the aerodynamics of wind turbines (see also Ch 7), the turbulence models in existing software appeared to be inadequate, as the turbulence around turbines covers a large array of length scales and is neither isotropic nor homogeneous. Other groups, like the Multiphase and Reactive Flows group of the TU Eindhoven, also developed software, in this case for combustion, which could be coupled to commercial and open-source CFD codes.

Since the early days, computer simulations of fluid flows had always raised doubts, especially if researchers did not, deliberately or otherwise, compare their numerical results with measurements from the real world. Or if simulation programs were used by people who had little knowledge of 'real flows'. The last issue was addressed by professor Hoogendoorn in his farewell speech in 1998. Part of this speech was about his views on the developments in CFD: "The 'ultimate' model with complete DNS [Direct Numerical Simulation] for all kinds of flows is still far away. Therefore, turbulence models needed for time-averaged equations will continue to play a role for a long time. This is a controversy. In top research do you solely aim at a complete and exact description, or should you improve incomplete but useful models? But alas, what is more challenging for top researchers than controversies about important questions in their field? I myself expect that models can be validated by means of DNS in the coming ten years. Validation is also required for the applications in technology. The 'trustworthiness' of much-used models is of decisive importance. It is therefore gratifying that the Burgers Centre will participate, in the context of ERCOFTAC (European Research Community on Flow, Turbulence and Combustion) in a validation program with so-called benchmarks. The interest in CFD in industry and the Grote Technologische Insituten [Big Technological Institutes] ... is big and means an important support for research. ... For the future I see an important growth for CFD. One point of concern for me: within some sectors of the industry the in-house expertise with regard to physical fluid mechanics threatens to disappear due to dismantling of R & D. This is dangerous: before you are aware of it you will make the wrong calculation." [translation by FA]

Doubts about computer simulations were also raised in the mind of professor Frans Nieuwstadt in 1999 when he was present as member of the committee during the defence of the PhD thesis of Arthur Petersen. Petersen has written about this 'incident' in his book on the value of simulations of the atmospheric boundary layer and of climate models

(Petersen (2012)). Some months before Petersen's defence the issue of the reliability of models had become a national debate in The Netherlands after one of the senior statisticians of the Netherlands National Institute for Public Health and the Environment had published an article in a national newspaper in which he warned that the Institute was leaning too heavily on its computer models.

"On June 7, 1999, I publicly defended my doctoral dissertation 'Convection and Chemistry in the Atmospheric Boundary Layer'. In this dissertation, the main body of which consisted of three journal articles based on computer simulation, I argued that one of the uncertainties in regional and global computer models of air quality was significantly smaller than was previously thought. Formerly, it was not known whether the influence of turbulence on chemical reactions in the

## PERFORMING NUMERICAL CALCULATIONS IN THE 1930S: BURGERS AND THE BLADES

In the early 1920s Burgers was asked by Werkspoor to help in improving the efficiency of the centrifugal pumps of the yet to be built Lely pumping station, near Medemblik (see also § 3.3.1 and 3.4.1). The idea was that giving the blades – which looked rather similar to aerofoils – the right contour, could not only increase the performance of the pumps but also put an end to suction problems. Werkspoor also wanted to have a method with which the pressure distribution along the blades could be calculated since this gives an indication about probable spots where cavitation problems can occur.

Burgers set out to apply the method of conformal transformation (see also § 6.3.1) to find a calculation procedure for the flow between the blades. This was developed from about 1925 but it seems that only in February 1929 were the first numerical calculations performed by some of his staff. Apparently, Burgers concluded that there was still room for improvement as in June 1930 he decided to give the blades a shape which show 'only a tiny deviation from a logarithmic spiral'. With his conformal transformation he could give the blades a shape for which the flow around it is simpler and already known. He also considered taking another direction: replacing the blade by a 'system of vortices' which was already done at that time in the lifting-line theory for wings of airplanes. Finally, he decided to take the route of the conformal transformation, a method he knew well and which produced faster numerical results.

At the end of 1930 Burgers and Van der Hegge Zijnen could present an impressive report about their project. In it Burgers explained the reason for the long duration of the project. It had become clear to him that he needed

four transformations to get the desired shape. "The search for these transformations required several months of almost continuous calculation labour. Though it is probable that in case of a repeat of such a calculation the goal will be reached faster, this remains an objection against the method of conformal mapping." Burgers also stressed the 'difficult pliability' of the method: even if the shape of the blades was changed only a little bit, all the calculations had to be done all over again. All the calculations had to be done with six decimals in order to get enough confidence about the usability of the transformations.

In September 1931 the pumps were ready and the installation in Medemblik was tested. But even thereafter Burgers came up with some additional calculations. In his final report of 1933 he explained that the goal of these calculations had been to give more insight into the flow around the 'nose' of the blades, but that the efficiency had been somewhat disappointing. What they did learn was that the pressure would become quite low around the nose. Later, during the operation of the pumps, cavitation was indeed observed at this location.

Had Burgers' calculations indeed led to better pumps? In fact, during testing the efficiency of the pumps appeared to be only 67%, whereas during model testing in the WL 87% had been found. Thanks to Burgers' findings Werkspoor discovered that the cause for this had to do with the 'huge dynamical diastolic pressures at the entrance of the fan' and they managed to design a new fan which indeed showed an efficiency of 87%, under all circumstances.

Burgers' calculation method has been used for several other centrifugal pumps at the WL. From about 1960 the computer took over the annoying parts of it.

atmospheric boundary layer could be neglected. I, together with my colleagues from the Institute for Marine and Atmospheric Research Utrecht (IMAU), using a hierarchy of computer models, had shown that this neglect was allowable. One of the opponents, Professor Frans Nieuwstadt of Delft University of Technology, sternly questioned me about the reliability of my research results until he was satisfied with my final answer, that I was confident about my research results only within a factor of two. His main problem with the work was that only simulation models of different complexity had been compared with each other, and no comparison had been made with experimental or observational data. My contention was that the most complex simulations that I had done using the national supercomputer of the Netherlands were more reliable for answering my research questions

than were any of the sparse experimental or observational results reported in the literature. This was judged by Professor Nieuwstadt to be a 'medieval position'. I disagreed since the large-eddy simulation (LES) model that I had used had been rigorously compared with experimental and observational data. The only thing I had done, I claimed, was to apply this model to a somewhat different problem, which was extremely difficult to approach experimentally or observationally. After this minor public controversy, the episode ended well since the doctorate was awarded by the committee without any objections." [references have been left out of this text]

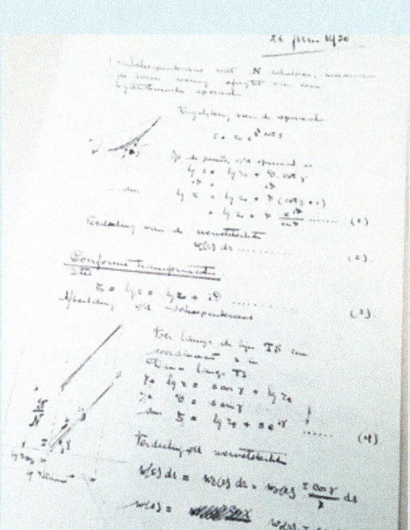
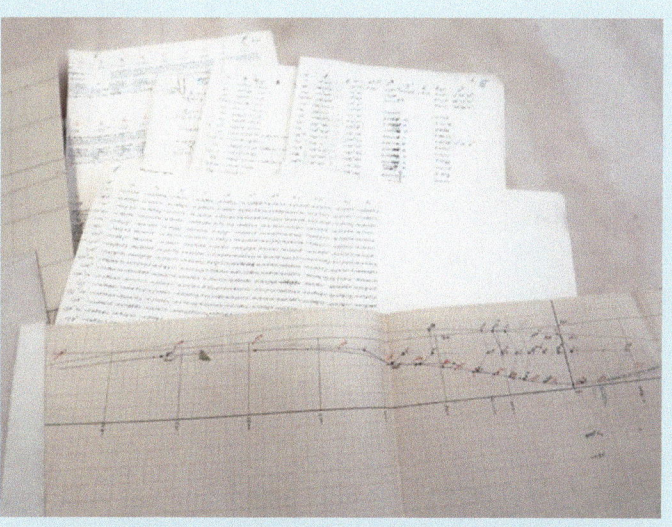

↑The Burgers Archives in Delft still contains loads of papers showing thousands of computed figures which were needed for the design of the impellers for the Lely Pumping Station. Van der Hegge Zijnen, Burgers' main assistant, must have spent hundreds of hours on the (mechanical) calculating machine. Burgers himself worked out the calculation method using conformal transformations (see also § 6.3.1). (courtesy of Burgers Archives / Delft University of Technology)

→The impellers with the blades which Burgers designed for the Lely Pumping Station in the 1930s have been used for many decades. They seem to have been replaced by fairly similar replicas in the 1980s (after which the cavitation problems continued to show up). The Hoogheemraadschap which is responsible for the station recently decided to install new fans, which are 'fish-friendly' and have a higher volume output. Radial fans have become rather outdated and many have been replaced by mixed-flow or axial fans. (courtesy of Leo Broers / Hoogheemraadschap Hollands Noorderkwartier)

© Springer Nature Switzerland AG 2019
F. Alkemade, *A Century of Fluid Mechanics in The Netherlands*, https://doi.org/10.1007/978-3-030-03586-0_7

# 07

## CAPITA SELECTA

Some topics from Dutch fluid mechanics,
not mentioned in the previous chapters,
deserve a separate description.

↑From about 1985 Hirschberg and co-workers of the TU Eindhoven did research on flow-induced pulsations in pipe systems with side branch cavities. Notice the video camera on the right. (courtesy of Mico Hirschberg / Eindhoven University of Technology)

↑In the 1970s TNO improved its analog PULSIM computer and validated it with field measurements. On the right a PDP-11 computer. Today the Fluid Dynamics group of TNO is still world-leading on subjects like pulsations/vibrations in pipe systems of reciprocating machines and flow-induced pulsations. From the 1990s the simulations could be validated with measurements from TNO's Flow Lab. (courtesy of TNO, Deaprtment of Heat Transfer and Fluid Dynamics)

# AERO-ACOUSTICS

Where sounds find their origin in the interaction of flowing gases with various structures, we speak of aero-acoustics. Sometimes these sounds are deliberately made and required, e.g., speech and singing or musical instruments. One of the first theories about how humans produce sounds with their voice was the so-called myoelastic-aerodynamic theory formulated in 1958 by Janwillem van den Berg, a lector in medical physics at the University of Groningen.

Most of the sounds which are studied by fluid mechanicists are unwelcome, and the purpose of the research is usually to find the origins of the sounds and thereby a means of reducing the noise level. Examples of sounds which have been studied by Dutch researchers are those produced by the blades of wind turbines, by flames, and by boilers.

The first location in the Netherlands where sound generated by flowing gases was studied was probably the NLR. The growth of aircraft traffic from the 1950s stimulated research into the reduction of the noise of airplanes. In the 1960s Zandbergen and others presented a theory of propeller noise. In the late 1970s NLR researchers started aero-acoustic experiments in one of the wind tunnels of the DNW (see § 6.2.2).

At the TU Eindhoven research on aero-acoustics has a long tradition. After the discovery of the huge Groningen gas field in 1959, pipeline transport became a common phenomenon in the Netherlands. In the 1980s Gasunie, owner of the Dutch gas transmission network, contacted Eindhoven. From about 1985 research was started on pulsations in gas pipelines with closed branches. This was done in the group of Rini van Dongen and of Mico Hirschberg. The latter started as a researcher at Shell and later became professor in both Eindhoven and Twente. The Eindhoven group also did research on the dynamic behaviour of volume flow meters. These flow meters can be disturbed when there are acoustical perturbations in the turbine of the flow meter. In such cases the measurements become unreliable. The interest in vibrations in pipes led to research on musical instruments. One of the discoveries done by Hirschberg and co-workers is the role which shockwaves play in the production of the typical 'copper sound' of trombones and similar wind instruments.

At TNO the simulation of pulsations in pipe systems started in 1965 when an engineer of the Dutch compressor manufacturer Thomassen approached the TNO institute TPD (see § 4.2.1). Compressors are notorious for causing pulsations. The PULSIM project (PULsation SIMulator) was initiated as a joint industry project in 1968. The sponsors were Dutch industries like BPM/Shell and Unilever, which felt the necessity of an improved method of vibration abatement. In 1970 the first calculations could be performed on the PULSIM analog computer, which was built by the TPD itself. After a try-out period TNO started to carry out pulsation studies on a regular basis and the need for sponsoring disappeared. Therefore in 1972 the formal relation between PULSIM and sponsors was terminated, and facilities

↑In the 1990s Hirschberg initiated research on the flow through organ pipes (vortex shedding, etc.). (courtesy of Mico Hirschberg / Eindhoven University of Technology)

↑The DelFly is a very small autonomous flapping wing (also called a Micro Air Vehicle or MAV), developed at the TU Delft and inspired by the flight of (fruit) flies. The first version was built in 2005 by a group of students. In 2008 the DelFly Micro was built, with a mass of 3.07 g and a wingspan of 10 cm! It could make flights of up to three minutes by controlling thrust, roll, and pitch. The DelFly II of 2013 was the very first MAV that could perform autonomous flight. Recently the Delft MAV Lab presented the DelFly Nimble which can hover or fly in any direction. It is controlled through insect-inspired adjustments of motion of its two pairs of flapping wings. Studying the behaviour of the DelFly has also increased insight into the secrets of the flight of flies. (courtesy of MAVLab – TU Delft; CC BY-SA 4.0 – www.delfly.nl)

have since been made available to others on a commercial basis. The PULSIM machine soon became known all over Europe. The calculation of the mechanical responses of the piping structure due to pulsations started in 1976, initiated because of a vibration problem in a gas compressor station of one of TNO's clients. In 1982 the first digital version of PULSIM was presented. Since then, this software tool for designers and engineers has been continuously improved. New models were developed and validated in contract research or joint industry projects, and some of these found their way to the engineering consultancy business (of which PULSIM has also become a part).

# ANIMALS

## FLYING

Jacob Jongbloed (1895–1974) of Utrecht University, an expert in the physiology of pilots, wrote about the aerodynamics of bird flight in 1938 but he seems to have been the only one in the Netherlands who was interested in this topic before the Second World War. Burgers got involved in the topic after the War when Everhard Slijper, professor of veterinary anatomy, asked for his help in writing a book which was published in 1950 under the title De vliegkunst in het dierenrijk (The art of flight in the animal kingdom).

After this it was about half a century before the flight of birds was again studied in the Netherlands. In Science in 2004 researchers from Groningen and Leiden (including biologist John Videler) published their research on the flight of the common swift. They had discovered that during the flight of this bird a long cylindrical vortex arises, the so-called leading-edge vortex. This had already been found on the wings of the Concorde airplane but not on those of birds. A model of the common swift wing was investigated in a water tunnel.

Research on animal flight was also performed in the last decade at the University of Wageningen, by David Lentink in the group of professor Johan van Leeuwen. They also studied the role of vortices and discovered that not only birds (like the common swift and the hummingbird) but also insects and maple keys (winged seeds or samaras) generate a vortex on top of their wings (which leads to a huge underpressure there) and that this vortex stays there. Lentink used high-speed cameras to study animals and maple keys in a vertical wind tunnel. In 2010 his team, which was called the Vliegkunstenaars (Flight artists), won the Academic Annual Prize for their proposal to train amateurs in using high-speed cameras (which they could borrow) and thereby collect hundreds of amateur videos of flying animals.

## SWIMMING

One very important discovery by a Dutch scientist related to the swimming of fish is not very well known in the fluid mechanics community. It is the discovery in the early 1960s by the physiologist and biologist Sven Dijkgraaf (1908–1995) of the functioning of the lateral line sensory systems of fish. For many years its functioning had remained an enigma and it was thought that the system was just there for secreting slime. Dijkgraaf showed that it is one of the most advanced flow sensing systems ever to have evolved, one with which fish can, e.g., sense objects in their vicinity thanks to the presence of a wake around these objects.

In the 1970s in de section of Theoretical Aerodynamics of Steketee in Delft some research was performed on the swimming of fish in waves. Dolphins can make an efficient use of the wave energy for their own transport. Quite recently researchers of the University of Groningen solved the so-called boxfish paradox: these rather peculiar fish were supposed to have a low drag resistance and could maintain their heading very well, but this was not in agreement with the way they behave near corals (low speed, making turns all the time). In Groningen a printed model of a boxfish (or trunkfish) was studied in a water channel and it appeared that its resistance was not as low as had been supposed and that it was not very course-stable.

## ATMOSPHERE

The fluid dynamics of the Earth's atmosphere was for a long time ill-understood. Only after the Second World War were attempts made to model the – turbulent – phenomena. Although concerns about air pollution would only become widespread in the late 1960s, in 1958 a Symposium on Atmospheric Diffusion and Air Pollution had already taken place in Oxford. Hinze contributed to it with a paper on heat transfer in the boundary layer.

Many aspects of atmospheric processes were still rather unknown up until around 1970; e.g., the influence of the Earth's surface on flows in the lower part of the atmosphere. In the late 1970s theory and observations of the (non-stationary and non-homogenous) turbulent stable atmospheric boundary-layer (SBL) started to develop. One of the researchers who contributed to progress in understanding the SBL was Frans Nieuwstadt (see § 4.1.1) who had started his career at the Physical Meteorology section of the KNMI in 1972. He introduced the concept of 'local scaling': the hypothesis that the (dimensionless) turbulence characteristics of the SBL can be expressed as functions of a single parameter in which the height above the Earth's surface is present. His local similarity theory has since found a lot of observational support. Nieuwstadt's second important contribution consisted of a series of nocturnal boundary-layer experiments in the late 1970s done at Cabauw (see § 6.2.6). These resulted in one of the few available data sets containing detailed mean and turbulence characteristics of the quasi-stationary SBL.

More recently research on the atmosphere has been carried

out in Eindhoven, by professor Bas van de Wiel and co-workers. They studied the relationship between the dynamics of turbulence at night and the formation of ground frost and related phenomena. They succeeded in developing an explanation of the impact that cloud coverage and wind speed have on the dynamics of near-surface turbulence and hence on near-surface temperatures. In Delft air pollution has been and still is a theme of research, e.g., in the DisTUrbE project: Dispersion in the Turbulent Urban Environment.

## COASTS

The morphological behaviour of the Dutch sea coast has been studied for more than one hundred years; the first thesis appeared in 1912. Johan van Veen (see § 3.4.1) discovered in the 1930s that the transport of sand near the coast was mainly determined by the sea surf flows. The NIOZ (see § 4.2) had a department for Physical and Chemical Oceanography from 1969 and a department for Sea Pollution from 1971 which shows the growing interest in this topic. Coastal and beach erosion was investigated experimentally with models at the WL. This kind of research requires huge basins since it is hardly possible to scale down sand grains. For more than twenty years the Netherlands Centre for Coastal Research (NCK) has been a cooperative network of private, governmental and independent research institutes and universities, all working in the field of coastal research. One of their research themes is 'Hydrodynamics'.

Coastal research became well-known with the general public thanks to the Zandmotor (Sand engine), a unique phenomenon worldwide. In 2011 a sandbar-shaped peninsula of about one square kilometre was deposited by dredgers at the coast of the province Zuid-Holland, south of The Hague. It is expected that these constantly evolving peninsulas will help to strengthen the coastline. In this way nature is used as partner of man in the constant fight against the deterioration of the coast due to wind and water flows. One of the effects which has been observed: sand grains from the peninsula are transported by the wind towards the coast and are deposited on the existing dunes or make new dunes. The Zandmotor method is expected to be more cost effective, and also helps nature by reducing the repeated disruption caused by replenishment. Evaluation of the Zandmotor in 2016 showed that it was behaving as expected. Researchers of the TU Delft study the currents around the peninsula, using dye and buoys with GPS equipment.

## COMBUSTION / FLAMES

Research on combustion in the Netherlands started at the Proefstation of Shell in Delft where processes in combustion

←A remarkable simulation of a swimming animal was done by FlowMotion, a consultancy bureau for heat transfer and fluid dynamics problems. FlowMotion was founded in 1996 in Delft and has since become an expert in using CFD to solve problems for a broad range of clients. Some years ago the Zeehondencrèche (Seal Creche) in the North of the Netherlands asked FlowMotion to simulate the flow around seals (i.e., seals with and without a transmitter on their body). The Zeehondencrèche studies the activities of seals in nature and can follow them by means of these transmitters. Though the method worked well, doubts had arisen about the influence of the transmitters on the swimming capability of the animals. The simulations showed that the energy efficiency of the animals decreased by about 12 per cent after a transmitter was fitted. They also showed that behind the transmitters a low shear stress was present; in this area adhesions had been found on severely emaciated seals. (courtesy of FlowMotion / Roy Mayer)

→In the 1970s TNO started research on environmental issues and established a Subcommittee on Air Pollution. One of the goals of this Subcommittee was the determination of a national model for the calculation of the spread of pollution, and to this end a workgroup was formed. Among its members were Frans Nieuwstadt on behalf of the KNMI, and researchers from KEMA, TNO, AKZO, and Shell. In their report of 1976 the workgroup advised using a Gaussian plume model. In a report published five years later by the Subcommittee advice was presented about a calculation method.

In Nieuwstadt's copy of this report one finds several (very) critical notes in the margin which shows that consensus had not yet been reached. (courtesy of Nieuwstadt Archives / TU Delft)

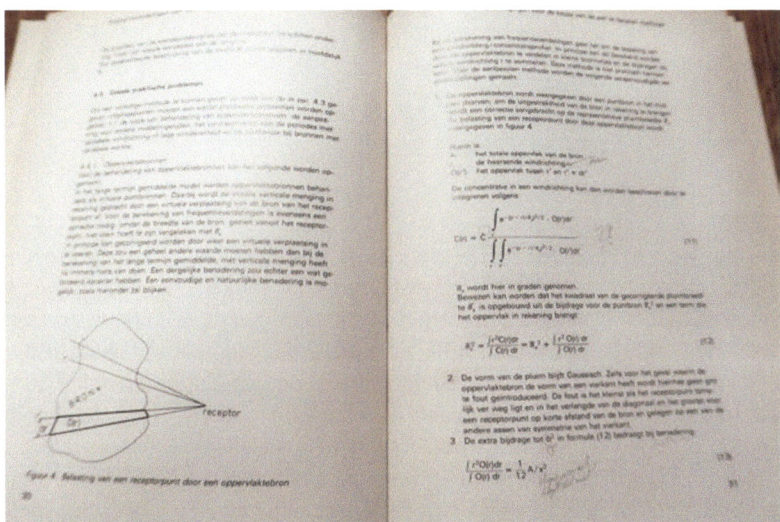

↑In the 1980s Nieuwstadt wrote some influential papers on the nocturnal atmospheric boundary layer. His paper published in the Journal of the Atmospheric Sciences in July 1984 led to a flow of reprint requests from all over the world. (courtesy of Nieuwstadt Archives / TU Delft)

↑The Zandmotor. (courtesy of Rijkswaterstaat / www.beeldbank.rws.nl)

↑In 1965 a new Aerodynamics Laboratory was put into operation at the IFRF premises in IJmuiden. (courtesy of IFRF / Lucy Straker)

↑Research on gas flames at the TU Eindhoven in 1997. (courtesy of TU Eindhoven archives)

engines were studied from 1928 (see § 3.3.2). However, the first location in the Netherlands where flames were studied systematically was not in Delft. From 1948 the International Flame Research Foundation (IFRF) was in operation in IJmuiden, the small town where the Hoogovens company (Royal Dutch Iron and Steel Company) was located. Hoogovens needed the IFRF to find answers for urgent questions about their switch to oil burners. A new furnace was built to do tests, primarily to investigate improvements to furnace efficiency. English and French researchers came to IJmuiden for advice and help. The IFRF was not a division of Hoogovens but the company was of course involved and provided money to the committee who organized the research. It was only in 1956 that the research facility achieved official status and was renamed Nederlandse Vereniging voor Vlamonderzoek (Netherlands Association for Flame Research). An international team of researchers was changed every few years.

Among the research topics of the IFRF was combustion aerodynamics, which became especially important in the 1960s. One of the purposes was to find a means for generating flames possessing distinct fluid flow characteristics. In one of the experiments the effect of swirl applied to the combustion air stream was investigated. In 1965 a new aerodynamics laboratory was opened and one of the researchers there was Alan Chesters, who would later work on two-phase flows at the universities of Delft and Eindhoven. One of the new burners developed in IJmuiden, around 1980, was the Aerodynamically Air Staged Burner which met the increasingly stringent rules on NOx emissions. This burner had a central coal injector and a swirling combustion air stream.
One of the Dutch advisers of the IFRF was professor Hoogendoorn from Delft (see § 4.1.1). His research on efficient burners started soon after his appointment in 1971, a result

of the growing concerns about environmental pollution and energy efficiency. Thanks to their knowledge about turbulent flows and mixing, Hoogendoorn and his co-workers were able to design better burners. Experiments with LDA led to their famous 'Delft double jet burner', one of the few burners with a generally recognized benchmark validation value. The fuel used was methane which is much more relevant for the design of natural gas burners than hydrogen. The Delft burner was depicted on the front cover of Hoogendoorn's farewell lecture of 1998 in which he proudly claimed that his team was now able to perform a reliable numerical simulation of an oven; though the modelling of the turbulence was still difficult...
As for many types of complex flows in which chemistry is involved, numerical simulations have largely improved. At the Multiphase and Reactive Flows group of the TU Eindhoven (with professor Philip de Goey), combustion has also been one of the main topics. As is the case with several other groups related to fluid mechanics, this group has developed software which they offer on their website to commercial users: "Simulations employing detailed chemistry and transport models are used to investigate the structure, emissions, dynamics, and stability of laminar flames. Our FGM [Flamelet Generated Manifolds; FA] model is extended to account for heat loss, fuel stratification, preferential diffusion, flame stretch/curvature effects, pollutant formation (UHC, CO, NOx, soot), and ignition/extinction. The interaction of flow and chemistry in turbulent flames is unravelled by using an in-house developed high-fidelity DNS code. Turbulent combustion models are developed and tested in an in-house LES code and validated against lab scale experiments. The knowledge from these studies is translated into efficient and accurate FGM-based models that are coupled to commercial and open-source CFD codes for the simulation of combustion in engineering applications."

# HUMAN BODY

Adriaan Moens (1846–1891) was a Dutch physician and physiologist. He became assistant to a professor of physiology in Leiden and there he began his work on arterial wave travel using reservoirs, elastic tubes, and air chambers. These studies formed the basis of his doctorate in 1877. The key finding of his work was an empirical relationship that described the velocity of pulse propagation in elastic tubes (e.g., blood vessels). Except for a numerical constant this turned out to be identical to the theoretical prediction derived by Korteweg (see § 2.4) in 1878, and the relationship is now known as the Moens–Korteweg equation.

During the decades thereafter, not much seems to have happened at the intersection of fluid mechanics and medical sciences in the Netherlands; only at the Laboratory of Physiology and the Laboratory for Medical Physics in Amsterdam was some research done. From 1977 this discipline was called bio-fluid mechanics, the section of fluid mechanics where flows inside the human body are studied. The first activities within the fluid mechanics communities seem to have taken place in Eindhoven in the 1970s, where attempts were made, by Van Dongen and co-workers, to make physical models which could explain the early closure of the aortic valve during flow deceleration. One of the PhD students in this field was Anton van Steenhoven, who had started as a student in 1969 and became professor of Energy Technology in 1990.

The studies in Eindhoven on flow phenomena related to the blood flow in arteries (usually laminar) and the heart and led to an interest in the visco-elastic behaviour of blood and tubes. In the 1980s both in-vivo and in-vitro experiments were performed, and this led to the development of fluid-structure interaction models in the 1990s. Another PhD student in Eindhoven working in this field was Frans van de Vosse (Numerical analysis of carotid artery flow, 1987) who is now professor of Cardiovascular Biomechanics at the TUE. One of the research themes of his group is 'Vessels under stress'. The aim is to develop clinical measurement techniques to assess the mechanical and morphological properties of the arterial wall and to use computational models to predict the development of vascular disease (atherosclerosis, aneurysms) and the outcome of clinical interventions.

From the early 1990s the CFD group of the University of Groningen had a section for 'computational hemodynamics', which collaborated with medical researchers at the University Medical Centre of the same city. Initially they solved the Navier-Stokes equations for the blood flows using the SEPRAN finite element package. SEPRAN was sold from the early 1980s, based on an earlier package developed at the TU Delft. Later the ComFlo package (see § 6.3) was used.

Saša Kenjereš of the Transport Phenomena section of the TU Delft and co-workers did studies on phenomena related to diseases which have to do with blood flow. Some years ago they performed PIV measurements of pulsating flows in reproduced parts of the vascular systems of real patients. More recently they studied methods to predict so-called hypoxic regions in the cerebrovascular system, which regions have been associated with the emergence of dementia. At the University of Utrecht PhD research was done on 'micro swimmers': microscopic particles which can be used for several medical purposes such as the delivery of medicines.

The Biomedical Flows group at the University of Twente is mainly working on the different applications of bubbles in the biomedical field. Coated microbubbles are used in ultrasound imaging to enhance the contrast in cardiac or liver perfusion images (the Erasmus Medical Centre in Rotterdam has also been active in this field). Bubbles can be targeted to specific cells for molecular imaging to non-invasively detect the presence and location of diseases such as cancer or atherosclerosis. Furthermore, the bubbles can be exploited to generate acoustic streaming and jetting near cell boundaries which leads to permeation, destruction or removal of target cells.

Other flow phenomena in the human body seem to have got much less attention in the fluid mechanics community. In 1903 a PhD thesis was defended at the University of Utrecht, entitled Bijdragen tot de aërodynamica der luchtwegen (Contributions to the aerodynamics of the respiratory tracts). From about 1990 theoretical and numerical research was done on the aerodynamics of the lungs in the group of professor Steketee in Delft, in collaboration with the Academic Medical Centre in Amsterdam. This study was related to the idea that the delay in the unfolding and development of the lung tissue in new born babies could be reduced by modulating the artificial respiration with a periodic pressure perturbation. A PhD student discovered that lung researchers had been using a wrong lung model. He developed a new convection-diffusion equation with which the optimal respiration for a patient could be determined.

The group of Kenjereš in Delft, mentioned above, performed Large Eddy Simulations of an aerosol distribution (a suspension of tiny droplets in air) within the human respiratory system. They proved that the magnetic steering of aerosols towards the left or right part of the lungs was possible.

# MICROFLUIDICS

Microfluidics is one of the youngest branches of fluid mechanics and started to mature in the 1990s. In the Netherlands it got broad publicity when in 2009 professor Albert van den Berg of the University of Twente received the NWO Spinoza Prize (the highest award in Dutch science). This was for his key breakthroughs in the understanding and manipulation of fluids in micro- and nanochannels, and for the application of this knowledge to areas such as new medical equipment. After Van den Berg's pioneering work, many other groups started studying and developing microfluidic flow, devices, and applications. A prime example is the so-called lab-on-a-chip: a miniature laboratory which fits completely onto a single chip. These tiny

# FOODS

## UNILEVER AND FLUID MECHANICS
### JO JANSSEN, UNILEVER RESEARCH AND DEVELOPMENT, VLAARDINGEN

Unilever has a wide product portfolio. At the 'dry' side we can distinguish a range of powder/particulate mixes, which may be granulated to reduce dustiness and allow more easy and reliable dosing. The 'wet' side ranges from structured liquids via pasty materials to semi-solids. From a scientific point of view, the recipes and microstructures are complex, and far away from academic model systems. The product rheology is usually evolving during the manufacturing process, for instance due to changes in particle/droplet/bubble size distribution, or phase transitions like crystallization or dissolution of some components.

Taking the optimization and scale-up of such processes beyond a purely empirical level obviously requires adequate knowledge of multiphase flow, kinetic transport phenomena (heat and mass transfer), and phase transitions. Computational tools are increasingly used to save time and resources by being '1st time close'. This can go even up to the prediction of consumer-perceived properties like 'creaminess' and 'thickness'. Such properties cannot be defined in a strict scientific way, like viscosity and yield stress, but can be correlated to measurable physical parameters. This involves the use of trained panels, which characterize the products in a reproducible way on a dedicated score card (e.g., a 1 to 10 scale). An example of such a 'from ingredients to consumers' simulation can be found in a fairly recent, Unilever-sponsored PhD thesis on the modelling of mayonnaise production [1].

### SPREADS PROCESSING

Unilever was founded as a merger between the Margarine Unie and Lever Brothers, so it is no surprise that the production of margarines and other fat spreads is one of the most thoroughly researched processes. For decades the dominant process has been the so-called Votator© line. It comprises one or more scraped-surface heat exchangers (SSHE) and at least one pin-stirrer. In margarine processing, a warm and coarse pre-emulsion is pasteurized in-line, followed by rapidly cooling and emulsification in the SSHEs. The latter are cooled from the outside with boiling ammonia (-40 °C). Fat from the pre-emulsion thus crystallizes almost immediately at the cooled walls, but before this layer can reach any substantial thickness it is removed and mixed back into the emulsion by the scraper blades on the rotor. The overall result is that part of the fat solidifies in the SSHEs in the form of tiny (clusters of) crystals, which stabilize the water droplets (Pickering stabilization). The remainder of the fat crystallizes during kneading in the pin stirrer, or even during the first hours of storage in the tub, and forms a space-filling network [2]. The overall process is a combination of emulsification and fat crystallization, coupled via the effective emulsion viscosity and the Pickering stabilization of the droplets. Droplet size and solid fat content change substantially along the production line.

Unilever R&D Vlaardingen (URDV) has contributed significantly to the understanding and modelling of SSHEs, which are also used in the manufacturing of ice cream. For instance, researchers have visualized and characterized the flow profiles and residence time distribution for Votator-type SSHEs [3]. They used a partly transparent (Perspex) unit, and visualized streamlines by injecting dyes into model liquids like water-glycerol mixtures. Flow regimes were characterized in terms of Reynolds and Taylor numbers. Heat transfer in SSHEs for margarine-type emulsions has been studied as well.

Within Unilever, this work has systematically been extended and combined with models for fat crystallization, resulting in a simulation tool for heat transfer and solid-fat formation in margarine processing. This allows 'in-silico' recipe testing, process optimization and trouble-shooting.

Despite the successful use of Votators, URDV has continued scouting for alternative technologies with better performance on sustainability and on the handling of challenging recipes. In 2012, products made via 'cool blending' were introduced to the market. The basic concept of this process is simple: rather than heating and cooling a complete pre-emulsion to crystallize only about 5 – 15 % w/w (mass fraction) of fat (on product), the fat is crystallized separately into a very fine powder. This powder is subsequently stirred into cold oil to make a slurry, into which the cold water is then emulsified.

Successful implementation of this concept depends on having a technology to make a suitable fat powder. Conventional spray crystallization does not work, but a specific version known as Particles from Gas-Saturated Solutions (PGSS) does. In PGSS a melt is saturated with dissolved $CO_2$ at high pressure and then sprayed over a nozzle. The rapid nucleation and expansion of $CO_2$ bubbles in the mixture upon depressurization strongly enhances the 'atomization' of the liquid. Moreover, the $CO_2$ expansion provides nearly instantaneous cooling (Joule-Thomson effect). The result is a very fine fat powder, which turned out to have the required properties for use in spreads production via cool blending.

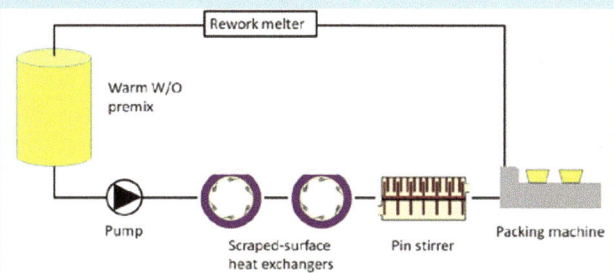

↑Typical Votator© line for spreads production. (courtesy of Unilever)

↑Viennetta ice cream. Contrary to what many people expect, the wavy pattern is due to a hydrodynamic instability in the flow from a stationary rectangular nozzle, rather than being produced by a sinusoidal nozzle motion (several movies can be found on YouTube). (courtesy of Unilever / photo by Fons Alkemade)

↑Illustration of the wide range of consumer products made by Unilever. Margarines/spreads have been a major example during the period covered by the present book, but Unilever sold this branch in 2018. (courtesy of Unilever)

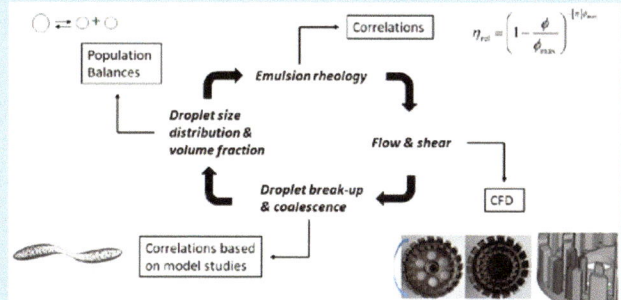

↑Interrelated processes in emulsification modelling. (courtesy of Unilever)

The first steps were taken in a STW project at the TU Delft [4], supported amongst others by Unilever. The aim of that project was to develop a continuous PGSS version for triglyceride fats. After completion of the PhD project, the process has been scaled-up within Unilever. The industrial scale version is unique in the world.

### EMULSIFICATION

Mayonnaise and spreads are examples of emulsions for which the droplet size distribution has a substantial effect on the product properties. Emulsification processes and their modelling have for long been a key interest for Unilever. While a lot of academic studies involve emulsification by turbulent flow, Unilever typically faces viscous emulsions with non-Newtonian continuous phases (fat crystal slurries, solutions of starches and/or other biopolymers). The emulsion rheology is complex and evolves during the process due to shear and droplet break-up/coalescence.

In the 1990s, a series of successive EU projects aimed to capture this in a pragmatic way by multiphase CFD. The basic approach was a specific version of the 'moments of distribution' approach in Population Balance Modelling. URDV participated in this work, focussing on experimentally validated models for droplet break-up and coalescence, in a form suitable for the chosen computational approach [5]. In the past, computing power and the stability of the simulations for realistic equipment geometries and conditions were often the key bottlenecks. Nowadays the limitation shifts towards the product rheology. Which constitutive equations and boundary conditions are appropriate for realistic CFD work on rheologically complex consumer products? Getting more insight into such questions has been a motivation to participate in the recently started NWO/ISPT project Controlling Multiphase Flow.

143

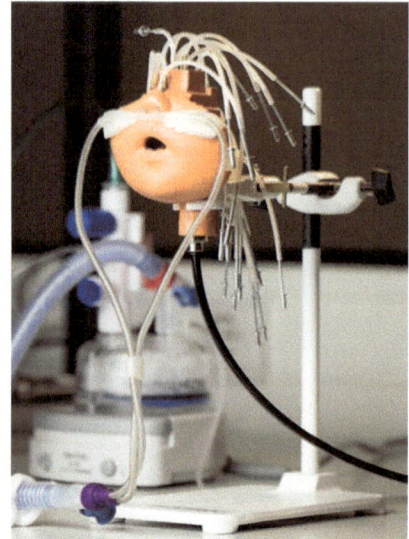

↑Fluid mechanicists at the University of Twente, together with medical researchers from the University of Groningen, are involved in research on the so-called nasal high flow therapy. NHF increases airway pressure and induces wash-out. Both effects are hard to assess in-vivo and large uncertainties exist. Therefore 3D-prints of faces and upper airways of humans are used to measure pressures and wash-out. (courtesy of Rob Hagmeijer / UT; photo by Rikkert Harink)

↑Jaap den Toonder is one of the leading scientists in microfluidics in the Netherlands. He started in this field at Philips and from 2004 he lectured on this at the TU Eindhoven. Here he shows a microfluidic chip with artificial cilia which can be used for active mixing. (courtesy of TU Eindhoven archives / photo by Bart van Overbeeke)

laboratories can, for example, be used in healthcare to perform rapid and cheap analyses. With insights into the physical characteristics of fluids in very small structures, electrical fields, magnetic fields, surface tension, or acoustic waves can be used to guide the movement of fluids in the channels on such a chip. Also, several (small) companies in the Netherlands are involved in making microfluidic products, such as in the High-Tech Factory on the University of Twente campus.

# SHIPS

Real breakthroughs in the design of ships are rather scarce. The Netherlands has been the scene, and sometimes the origin, of some more or less successful innovations.
In 1968 the world's first full-scale SWATH vessel was built in Bolnes, near Rotterdam. The "Duplus" was a so-called diving support ship, used for drilling seabed samples and as a standby vessel. SWATH is the acronym for Small Waterplane Area Twin Hull, a twin-hull ship design that minimizes hull cross section area at the sea's surface. Minimizing the ship's volume near the surface of the sea, where wave energy is located, maximizes a vessel's stability, even in high seas and at high speeds. After two years the "Duplus" was rebuilt into a hybrid form between SWATH ship and catamaran. The SWATH concept has continued to be the object of ongoing research and several kinds of ships have since been built on this basis.
Research on (innovative) ship hulls became a specialty of the Ship Hydromechanics Lab of the Technical University in Delft (see § 4.1.1). From 1970 up until 2009 measurements on seventy hulls were performed leading to the famous Delft Systematic Yacht Hull Series (DSYHS), which was started by Gerritsma and became the life work of Lex Keuning. Since 2010 all generated data has been publicly available, allowing for the validation of CFD codes.
One really successful innovation for which the Delft researchers can be credited was the axe bow. Since the eighties, Damen Shipyards and the TUD cooperated in a research programme

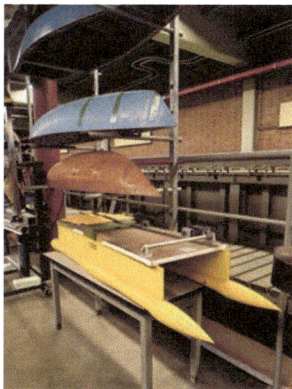

↑A model of a SWATH vessel is kept by the Ship Laboratory of the TU Delft. Above it some models are stored which are related to the Delft Systematic Yacht Hull Series.

↑A ship model with an axe bow in the pond around the building where the TUD Ship Hydromechanics Laboratory is located.

↑Innovation in ship design is also initiated by the industry itself. The shipbuilding company De Waal, a family business since 1938, developed the Easyflow rudder system in the 2010s. With the wheel straight ahead, the rudders are kept just outside the propeller wash, which flows much faster along the inside face of the rudder than along the outside. This creates considerable extra pressure on the inside, and lower pressure on the outside. The patented Easyflow aerofoil converts this pressure difference into extra thrust. Ships with this system have a higher fuel efficiency and produce less noise. (courtesy of De Waal bv)

↑Recently the aerodynamics of cyclists has been studied outside wind tunnels. In the Ring of Fire large-scale stereoscopic PIV is used to measure the velocity around the cyclist and to derive the drag forces. (courtesy of Alexander Spoelstra/ TUD; photo by Emile van Aelst)

aimed at improving the seakeeping characteristics of high-speed vessels. In the nineties, this cooperation produced the so-called Enlarged Ship Concept which led at the beginning of this century to the Axe Bow Concept, a wave-piercing type of ship's bow, characterised by a vertical stem and a relatively long and narrow entry. Based on this concept, Damen developed the Sea Axe patrol vessels and fast crew suppliers. The new bow not only improved operational ability and increased the comfort and safety of the crew, it also led to a significant reduction in fuel consumption and lower emissions.

# SPORTS

## SKATING

Skating – a sport in which Dutch athletes have for decades been very successful – seems to be the first sport which has been seriously studied in the field of fluid mechanics in the

Netherlands. After a first report on the possibilities of drag reduction for skaters in 1976, Dobbinga and others started experiments on cylinders and 'artificial legs' in the Low Speed Wind tunnel Laboratory of the TU Delft (see § 6.2.2) in 1987. With 'disturbing wires' and later with 'zigzag stripes' on these objects they tried to disturb the boundary layer with a positive result. The stripes, which had already been used on glider planes, appeared to reduce the drag by 15 per cent.
In 2014 Hoeijmakers and co-workers at the University of Twente tested the influence of twenty different fabrics for the skating suits of the national team in their wind tunnel. Furthermore, at the NLR skaters have been put inside a wind tunnel, as has also been done with a bobsleigh and its crew.

## CYCLING

The earliest known test of a professional cyclist in a wind tunnel is from 2007 when Olympic champion Leontien van Moorsel was observed in the wind tunnel of TNO in Apeldoorn

↑Attendees of the International Conference of Tank Superintendents in 1933 among whom are Burgers (15), Troost (5), and Van
Lammeren (21). (courtesy of MARIN)

↑A cavitating propeller in a cavitation tunnel (courtesy of MARIN).

↑Current research is moving from oil & gas platforms to floating
islands and floating wind turbines (courtesy of MARIN).

## RESEARCH ON SHIP HYDRODYNAMICS IN THE NETHERLANDS
### HENK PRINS, MARIN RESEARCH & DEVELOPMENT

From 1800 onwards, studies were conducted in France and
England on the performance of ships. The French approached
the topic theoretically, in line with their mathematical tradition. The English used a more pragmatic approach and tried
to understand differences between ships with scale models
and tug-tests. Soon they discovered that a ship that was
better in towing test was not always better in practice. As a
result, the tests had very limited value and the design of ships
remained a craft.

It is due to William Froude that model tests became valuable
after all. He invented the decoupling of wave resistance and
frictional resistance, and this allowed him to use model tests
to predict the full-scale speed of ships. When Froude demonstrated the correctness of his method in 1873, Bruno Tideman
took the initiative to build a towing tank at the Navy Yard in
Amsterdam. This tank was the second facility in the world.
Initially, it was mainly used for investigating Navy ships, but
soon it was also deployed for private ships, such as the yacht
Livadia for the Tsar of Russia. Unfortunately, the facility was
not used anymore after the death of Tideman in 1883.

While in the Netherlands research in ship hydrodynamics
came to a standstill, more and more towing tanks were built
in the surrounding countries. Therefore, around 1910 an
initiative was taken to again establish a Dutch towing tank.
This initiative was mainly pushed by the naval architecture
students' association 'William Froude' at the Technische
Hogeschool in Delft (now TU Delft). Unfortunately, Dutch
industry did not believe in setting up a Dutch towing tank,
and the initiative lost momentum. A new initiative from the
Dutch government failed around 1920 due to the economic
recession. In 1930, the Dutch government adopted a law to
stimulate applied research. Following this, TNO was founded.
Finally, in 1932 the Dutch shipbuilding test station NSP was
founded, later renamed Maritime Research Institute Nederland (MARIN).

A year after its foundation, NSP and the TH Delft organized
an international conference for the directors of towing tanks.
This conference was continued as the International Conference of Tank Superintendents, and still exists as the International Towing Tank Conference (ITTC). At the first conference
one of the attendees was Jan Burgers.

Over the years, the facilities of the NSP have been extended
with various cavitation tunnels, a towing tank for seakeeping of ships, a towing tank for inland waterways, a tank for
stationary offshore constructions in currents and waves,
and a depressurised tank. In addition to these experimental
facilities, early investments were made in numerical equipment. In 1960, the NSP opened a Computer Centre to make

drawings and calculations for shipyards. And in 1970, the first bridge simulator was opened to investigate the safety of ship manoeuvres (including the human factor).

MARIN received a significant impulse at the end of the 20th century, when the Dutch government made a large investment in new facilities. MARIN now has the most modern seakeeping & manoeuvring basins and offshore basin in the world, and has become the world leader in research on ship hydrodynamics.

In 2016, TU Delft, MARIN, and the maritime sector laid down a common vision for the future of maritime research. We expect that the sector will need to focus on zero-emission transport, vessels sailing autonomously, renewable energy extraction at sea, and floating islands. Enough challenges for the future!

## PROPELLER DESIGN:
## THE WAGENINGEN B-SERIES

The concept of a marine propeller is essentially already centuries old, as the ancient Greeks introduced the Archimedes screw. The application of this concept only started for ships when the steam engine had become practical. However, in the 19th century shipbuilders still had no idea of the influence of the form of the propeller on performance, placing ever stronger engines in ships to reach ever higher speeds. This continued until 1894, when the new Navy ship HMS Darling was delivered. This ship appeared to fail completely in reaching the predicted speed. This proved to be due to large-scale cavitation: the evaporation of water through the low pressure on one side of the propeller. That was the starting point for systematic research into propeller design and cavitation; research that is still ongoing today.

At the beginning of the 20th century all over the world special cavitation tunnels were built. In the Netherlands this was done at MARIN and at the TH Delft. In these tunnels one can examine whether a designed propeller can provide the required thrust and whether harmful cavitation will occur. But there was still little insight into which propellers would perform well and which would not. That is why MARIN started to set up a systematic series of propellers: the Wageningen series. The A-series had already been tested before the Second World War. This series consisted of propellers that had a profile over the entire diameter that looked like a wing. It soon turned out that these propellers did not perform well under heavy load. Therefore, after the war, a new series of propellers – the B-series - was started, which were formed differently around the tip of the propeller blade. Until the end of the 1960s, propellers were added to the series. The results were published in standardized diagrams. Later, these diagrams became available in many

design programs and today they are even available in online apps. The Wageningen B-series is still very important for both educational purposes and for propeller design in the maritime industry.

Although the series is named after Wageningen, there is also a strong link with the TU Delft. Successive professors of shipbuilding (Van Lammeren, Van Manen, Kuiper) also worked for MARIN and have contributed, with their students, to the creation of the series.

## OFFSHORE PLATFORMS: DRIFT FORCES

Since the beginning of the 20th century oil has been produced at sea. In the beginning, simple wooden constructions were used that were close to the beach. Later the constructions were made stronger by using steel and concrete. These structures were resistant to wave forces and could therefore be put further from the shore. Examples of this can still be seen in the North Sea.

Due to the increasing demand for oil, new oil fields had to be explored that were in deeper water. In deep water it is no longer convenient to put a fixed structure on the bottom of the sea. Floating systems were designed; some in the shape of a ship, but also some very new constructions were conceived.

The disadvantage of floating platforms is that they move up and down, and back and forth when waves pass by. This is especially the case in long (large amplitude) waves, a platform moves along with the waves; in short (small amplitude) waves the platform hardly moves. In the design of mooring lines, anchors, and flexible oil hoses this must be taken into account.

In the course of the 1970s, it was discovered that another problem occurs with waves: platforms want to drift off. We all know this from everyday life: if a football is in a pond, you can make waves in the water and drive the ball with the waves towards the other side. The ball does not only move up and down with the waves, but also drifts away. This also happens on a larger scale with oil platforms. This has great consequences for the mooring lines, but also for ships that come to offload the oil and transport it to shore.

In order to be able to take this drifting into account during the design of ships and platforms, new measurement and calculation techniques had to be developed. This has resulted in a long-term cooperation between TU Delft, University of Twente and MARIN. At MARIN, Johan Wichers worked on measuring drift forces. That was a cumbersome task in itself, as the ship model would drift away during the measurements.

The ship had to be kept in place with soft springs. But the springs influence the movements of the ship and thus the drift forces. Jo Pinkster, who later became professor of ship hydrodynamics in Delft, worked on a calculation method: based on the assumption that the ship makes small movements and that the waves are small, he could make an estimation of the drift forces in regular waves.

Because the assumptions of Pinkster were too restrictive for more extreme conditions, the University of Twente, under the guidance of professor Zandbergen, worked on a new calculation method based on non-linear waves and flows. From the mid-1980s to the end of the 20th century, several PhD students tried to perfect the method. Unfortunately, it turned out that the method was computationally too intensive to be practically applicable.

At the same time, at the TU Delft professor Hermans followed an alternative track. His researchers attempted to tackle the effects of current in the water and irregular waves. This line of research also suffered from great computation times; however, the results were shown to be useful and are included in the standard programmes currently used in Industry.

Since the beginning of the 21st century the subject has become also relevant for sailing ships. Ships were always built with an oversized engine. As a result, there was always extra power on the ship for emergency situations. But this also caused a ship to be more polluting than was strictly necessary. That is why international agreements have been made to stop over-dimensioning engines. This means that we need to study more carefully what engine power would be required in an emergency situation (called 'safe return to port'). A small drift force can already be decisive in determining whether a course can be held or that the ship runs out of its rudder and drifts away. The calculation techniques developed at the TU Delft are now also being used for this problem. Special validation measurements have been carried out at MARIN by Reint Dallinga.

With the rise of CFD (Computational Fluid Dynamics) it is now possible to bring all desired effects into the calculation of drift forces. But the calculation time is often still in the order of days for one single calculation. Since ship designers want to study a lot of different wave heights, wavelengths, and wave directions, a computational study will soon take months of calculation time. That is why the industry is still relying on the fundamental work done in the last century.

(see § 4.2.1). In 2016 the famous Dutch cyclist Tom Dumoulin did not have to enter one of the wind tunnels of the TU Delft himself but was replaced by a 3D printed mannequin. By means of PIV with helium-filled soap bubbles, staff of the TU Delft Sports Engineering Institute investigated the influence of the fabrics of the clothing on the drag forces.

The new Atmospheric Boundary Layer Wind Tunnel of the TU Eindhoven (see § 6.2.2) allows not only the study of flows in the built environment on a model scale but also full-scale studies of, e.g., a line-up of up to nine cyclists at short distances behind each other. Measurements together with CFD simulations allowed the group of Bert Blocken and co-workers to draw conclusions about the optimal sequence of the cyclists with regard to drag forces. One of the conclusions: the cyclist at the front also experiences drag reduction.

### SWIMMING

In a swimming pool in Eindhoven – well-known as a training facility for Dutch professional swimmers and a location for (international) swimming contests – a remarkable 'laboratory' was realized in 2016. From tubes integrated with the bottom of the training basin air bubbles can rise. When a swimmer passes, the movements of these bubbles are disturbed. These disturbances are filmed by several cameras and can later be analysed.

One of the aspects of swimming which has been studied at both the universities of Delft and Eindhoven is the influence of the spreading of the fingers on the efficiency for freestyle swimming. Theoretical, experimental and numerical considerations confirmed the conclusion which had already been drawn by others: spreading the fingers at a certain level can result in an increase in efficiency (and may increase the chance of winning).

# WIND

## BUILDINGS

As has been related in § 3.3.1 and § 3.4.3 Burgers and his co-workers became involved in research on ventilation from about 1930. Their expertise with regard to the ventilation of traffic tunnels was maintained in the Laboratory of Aerodynamics and Hydrodynamics at least up until the 1960s.

The earliest Dutch publication on the ventilation of houses is from 1871 but research in this field seems to have started much later. In 1942 the first lecture notes on ventilation were published in Delft, and in 1949 the first issue of Verwarming en Ventilatie (heating and ventilation), a journal for professionals in the installation branch, appeared. As for the ventilation of buildings, Burgers was involved in a remarkable

Twan van Hooff did a PhD study on ventilation in 2012 at the TU Eindhoven. Here he is standing in a ventilated model room during the Dutch Design Week 2014. (courtesy of TU Eindhoven archives / photo by Bart van Overbeeke)

In the early 2000s former astronaut and TUD professor Wubbo Ockels (1946–2014) and co-workers started to develop an airborne wind energy system in which use was made of kite-like wings. The original 'laddermill' concept was developed into a system with just one large kite attached with a cable to a winch with a generator on the ground. In 2007, the first 20 kW Kitepower system demonstrated that the concept worked. The Kitepower company was founded in 2016 and developed a 100 kW system. The Netherlands is one of most active countries in airborne wind energy with at least five promising start-ups. (courtesy of Kitepower (Enevate BV) – www.kitepower.nl)

In 1937 measurements were done in Burgers' laboratory on models of the (simple) office buildings of the BPM on the island of Curaçao. There, employees had experienced fierce winds outside, which caused strong winds inside the buildings (which had no windows). (courtesy of Burgers Archives / TU Delft)

At the TU Eindhoven research on wind turbines was started by Paul Smulders around 1973, the year of the oil crisis in the Netherlands. One of the aims there was the development of a simple and efficient wind turbine for pumping water in developing countries. Around 1980 the researchers had realised a test field with several different types of windmills. (courtesy of TU Eindhoven archives)

Fig. 2.
Opstelling van het model 1 à 25.

Fig. 4.
Opstelling van het model 1 à 200.

On the turntable in Peutz's wind tunnel in Mook dozens of different projects have been investigated: tall flats and office buildings, railway stations, hospitals, and also bridges. (courtesy of Peutz / Simone Straten)

# SPECIAL FLUIDS: STEEL

## FLUID MECHANICS AND STEEL PRODUCTION
### TIM PEETERS, TATA STEEL - RESEARCH & DEVELOPMENT

Hoogovens, the predecessor of Tata Steel in IJmuiden, started its business in iron and steel making on September 20, 1918. Initially the company only produced so-called pig iron, but before World War II the site had evolved to include a steel plant and a hot rolling mill. R&D and Analytical Lab facilities started in 1947, followed by several decades of further growth and development. The European Coal and Steel Community, established in 1952, enabled strong international collaboration in R&D in the European steel industry. In later years the steel industry underwent several rationalizations, mergers and acquisitions, triggered by strong global competition. Today, Tata Steel's European R&D organization in The Netherlands employs approximately 350 people in The Netherlands.

Stimulated by developments in the USA, where large steel mills were supplanted by smaller and flexible 'Mini Mills', the R&D Department of Hoogovens started in 1988 with an initiative for alternative manufacturing methods. This involved the idea of developing new production methods where process steps could be skipped or integrated. This led to the development of a 'thin slab' casting installation integrated with a rolling mill, called 'gietwalsinstallatie' or Direct Sheet Plant (DSP). This strategic project strongly relied on large-scale research in the field of fluid dynamics for continuous casting and temperature control in the rolling process. In a conventional continuous caster, the casting speed of the solid material lies between 1 and 2 m/min. In a thin slab caster, the total throughput of the installation is similar to a conventional caster, but the product dimension is much thinner and therefore the casting speed can be as high as 5 to 6 m/min (or 10 cm/s). This puts heavy demands on stable process control, because there is a higher risk of so-called break-outs. Experimental research was carried out by means of water models, using the fact that liquid steel has virtually the same kinematic viscosity as water. It was determined to what extent the high speeds in this process could lead to possible process disturbances. The technological solution consisted of the application of magnetic fields to dampen the turbulence and stabilize the flow in the casting mould. This groundbreaking experimental research was flanked by numerical simulations in which the influence of heat transfer and magnetohydrodynamics were included, which obviously was not possible in water modelling. In this way, a design was made in collaboration with international partners for an innovative casting machine. In 1998, it was decided to build the DSP, which was put into service in the year 2000.

In the 1980s a shortage of metallurgical coking coal led to higher raw materials costs, and R&D saw a new possibility of process integration through the so-called Cyclone Converter Furnace. This process aimed at the direct melting of iron ore without the manufacture of sinter, pellets or coke, so the traditional blast furnace would no longer be needed. The first step, namely the smelt cyclone, was successfully tested in the 1990s, but the upscaling to the second step,

project in Amsterdam during WWII: he was asked for advice on the improvement of the ventilation of the 'arrest cells' in the police headquarters. Burgers did some measurements on the spot.
Research on ventilation inside buildings was done by TNO from the 1960s and on ventilation inside greenhouses at the University of Wageningen from the 1970s. Ventilation became part of a field of engineering called 'bouwfysica' (building physics) which became one of the key research fields at the TU Eindhoven. Since the 2000s Bert Blocken, professor since 2011, has been involved in research on 'wind engineering' (and sports aerodynamics, see § 6.2.2) and has initiated research on ventilation and on wind loads on buildings.

The investigation of wind flows on and around the built environment has a rather long tradition. In 1931 the first measurements on a model house were performed in one of the wind tunnels of Burgers' laboratory. In the same year the RSL did measurements on the wind climate outside and inside a model of the cattle market building in Zwolle. In recent decades the TU Eindhoven has been the place where academic research was done. Applied research has been done for more than three decades in the wind tunnel of consultancy company Peutz. Peutz is one of the few independent firms of its kind with its own wind tunnel. For their clients their researchers investigate wind comfort, wind danger, pressures and loads, air quality, and the spreading of substances like smoke.

## ENERGY

As related in § 4.2 research on modern wind turbines started at ECN in the 1970s. From 1976 till 1986 they focussed on the Vertical Axis Turbine (VAT). They cooperated with the Fokker company which eventually designed its own VAT. From 1981 ECN also studied the Horizontal Axis Turbine (HAT) and for this they cooperated with the NLR. Though the investigations the converter, turned out - in 1998 — to be financially un-feasible. In 2004, however, a second component was added to this idea. As a result of the Kyoto Climate agreement, the European steel industry joined forces in the ULCOS (Ultra Low CO2 Steel making) consortium, aiming to prepare the European steel industry for 50% CO2 Reduction in 2050. At the time such a target seemed highly ambitious and still far away from practical implementation. Following the inventory of many new and existing ideas, the HIsarna Process was born. This process is a combination of the smelting cyclone developed by Hoogovens (and partners) with the High Intensity Smelting (HIsmelt) technology. The new technology was renamed HIsarna where 'Isarna' refers to the old-Celtic word for iron. The design and realization of this process is based mainly on complex CFD models. These models include turbulent multi-phase flows, particle separation in a cyclone, heterogeneous chemical reactions, heat radiation, and phase-transitions. Since 2011, Tata Steel IJmuiden has operated a pilot plant in which this process has been developed into a semi-continuous process in 2018, at a nominal rate of 5 to 8 ton/h. The HIsarna pilot plant capacity is only 1 to 2 per cent of a full-scale blast furnace, but is considered the minimum size for testing purposes. The HIsarna process has the potential to reduce costs in iron and steel manufacturing as well as to provide a solution for CO2 emission, because on the one hand the need for fossil carbon is lower and on the other hand all produced CO2 can be captured and stored very efficiently.

↑Physical water model of a typical continuous casting mould. Quantitative fluid dynamics measurements are conducted on bubble dispersion, flow stability, turbulence and 3D velocity fields in a 1:1 scale geometry (height 2 m, width 1.5 m, thickness 225 mm). (courtesy of Tata Steel)

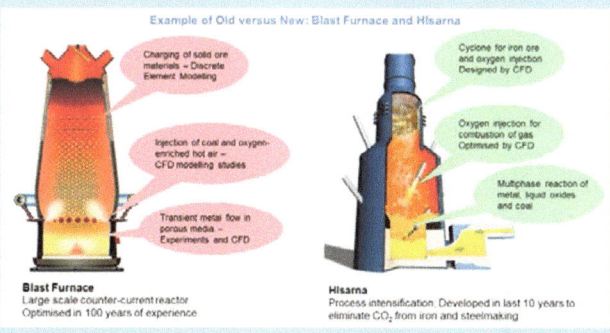

↑(courtesy of Tata Steel)

showed that VATs could certainly be viable, during the 1980s attention was completely directed towards the HATs. (In the 2010s researchers at the TU Eindhoven would revisit VATs.) From 1987 ECN studied wind turbines with flexible blades. With this system the rotor was better able to deal with the wind forces. Though this and other developments seemed promising, Dutch turbine builders (like Lagerweij/Lagerwey, founded in 1979) remained rather passive in innovation. Furthermore, some legislation in the Netherlands hindered their ambitions and the Dutch industry lost its lead with regard to flexible turbines in the 1990s (and Denmark took over). ECN continued its research and installed several full-scale turbines on its unique test field (see § 6.2.1). Today this field is deserted: the large turbines of today, which ECN would like to study, are not allowed in the dunes. ECN, and turbine builders, are now using a test field in the Wieringermeer, where they can also investigate the interaction between turbines in so-called wind farms.

Clusters of wind turbines had already been tested many years earlier at TNO Apeldoorn (see § 4.2.1) where model wind parks were placed in the wind tunnel. TNO produced the first software worldwide with which the effects of turbine clusters could be simulated.

Research on wind energy at the Delft University of Technology began around 1976 at the Institute for Wind Energy, starting with the tip vane project, an aerodynamic research project at the Department of Aerospace Engineering. Nowadays DUWIND is the wind energy network for research and education in Delft, bringing together six departments and thirteen groups. At the University of Twente numerical simulations of flows around blades and turbines have been performed since the early 2000s.

# SPECIAL FLUIDS: INK

## PRINTING FUTURE - DEFINING THE FUTURE OF INKJET PRINTING
HANS REINTEN, SENIOR SCIENTIST AT OCÉ,
A CANON COMPANY

Océ was founded in 1877, and since 2010 has been one of the main innovation centers for Canon, global leader in consumer and professional imaging. The company's vision lies in finding new ways to print information and images. We are focused on accelerating digital imaging technologies and developing high-tech printing products and services. The fundamental technology driving Océ's innovations is inkjet, and that requires an ever-greater understanding of fluid dynamics. Here, we outline our efforts in fluid dynamics and the resulting innovations in printing.

### THE GOLDEN AGE OF FLUID DYNAMICS
Fluid dynamics is one of the most vibrant disciplines in modern science. Thanks to the perseverance of research scientists and excellent partnerships, Océ is helping create a "golden age" of fluid dynamics. From drop formation to air bubble entrapment, from the behavior of complex fluids to turbulence, scientists are making significant progress in understanding the details of underlying mechanisms. Océ R&D specialists are applying this expertise to develop new generations of printing systems, suitable for an impressive array of applications.

### SUBSTRATES
Not only on paper, but also on "new" substrates such as glass, wood, and metal, inkjet technology has now firmly established itself as the primary digital printing method for applications as varied as retail display banners, books on demand, and interior design. And researchers at Océ have only just begun to explore its full potential.

Here are just a few of the creations possible with Océ inkjet technology, thanks to their understanding and application of fluid dynamics:
*Magazines, catalogs and books* - Océ has developed some of the world's best-selling digital inkjet systems that produce high-quality full-color prints on demand, on a range of low-cost paper types.

*Interior design* - Able to print on almost any material, including glass, wood, metal, ceramic tiles, laminates, and wallpaper, Océ digital printers are the dream of many interior designers.

*Elevated printing technology* - Perfect for fine-art recreations, elevated maps, signage, art reproductions, creative interiors, and educational materials for the visually disabled,

Océ elevated printing is creating exceptional products for both the consumer and the professional markets. It also enables industrial users to print moulds for quick samples or prototyping, as an alternative to vacuum forming and embossing.

### TECHNOLOGICAL BREAKTHROUGHS
Such products and applications have been made possible by decades of research in the field of fluid dynamics. Below are some of the many technological breakthroughs that Océ is applying in its latest innovations:
UVgel - Precision printing with UVgel inks.
High temperatures - Océ metaljet is the first technology for precision-jetting droplets of molten metal. It can be used with any conductive material with a melting point of up to 2000°C: pure metals, alloys, and molten semiconductors such as silicon. This opens up opportunities for printing jewelry, electronics, engine parts, and much more.
3D printing - Océ is currently exploring the value of their printing technology, piezo printheads, and 3D workflow software in industrial applications, where the biggest opportunity lies in low-cost multi-material Additive Manufacturing: 3D printing. Multi Material Jetting technology gives access to markets like biomedical, health, printing electronics, and many more.

### COOPERATION - BRIDGING THE GAP
### BETWEEN FUNDAMENTAL AND APPLIED KNOWLEDGE
In the past, academic and industrial researchers struggled to connect. A Dutch science foundation event in 2001 changed all that. It was here that leading researchers, Hans Reinten from Océ and Detlef Lohse from the University of Twente met. Their discussions marked the start of an important collaboration in the field of fluid dynamics. Océ continues to bridge the gap between applied and fundamental knowledge. In 2015, Océ Researcher, Herman Wijshoff was appointed Professor of Fluid Dynamics of Inkjet Printing at the Faculty of Mechanical Engineering at Eindhoven University of Technology. "Fundamental knowledge gives us a much better understanding of how droplets react on different surfaces and substrates," he says.
Océ is now using fundamental knowledge to gain a much better understanding of how physical characteristics change when droplets get smaller, for example. And how these characteristics affect the surrounding environment. The resulting model-based design uses accurate simulations and fewer prototypes to achieve success.

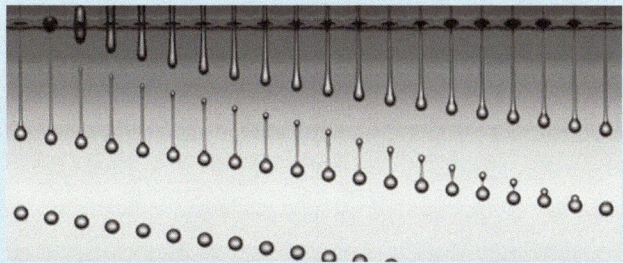

↑ Droplet formation from a 30 micrometer diameter nozzle of a piezo-acoustic inkjet printhead. Imaged using single flash strobo-scopic imaging with illumination by laser induced fluorescence. The temporal resolution is 600 nanoseconds. Nature picture of the year 2014. (Courtesy of Océ and Physics of Fluids group/UT)

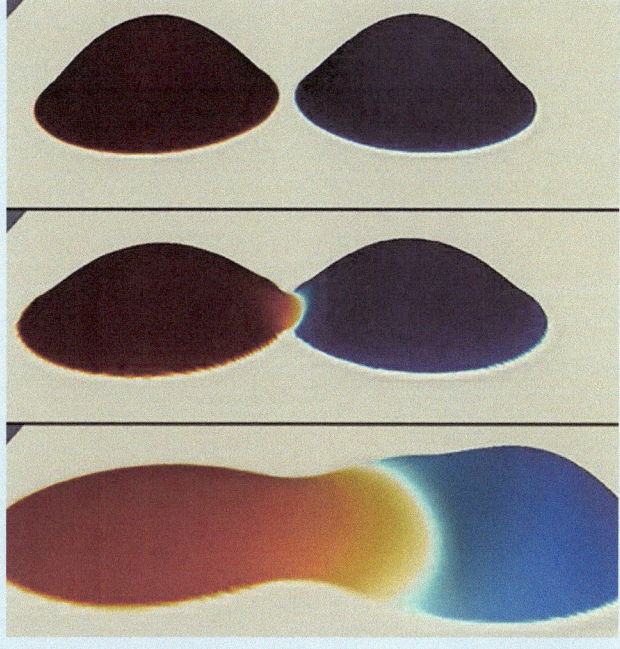

↑ Numerical simulation of the coalescence of different droplets showing the problem of colour bleeding. Due to the so-called Marangoni effect, the droplet with the lower surface tension in-vades and displaces the droplet with the higher surface tension. (Courtesy of Océ and Eindhoven University of Technology)

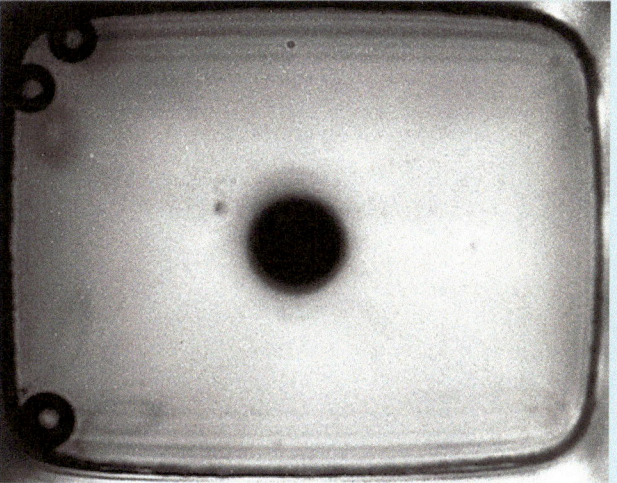

↑ Entrained air bubbles in a piezo-acoustic inkjet printhead, imaged using stroboscopic shortwave infrared imaging. (Courtesy of Océ and Physics of Fluids group/UT)

Nowadays Océ cooperates extensively with leading scientists at the forefront of fluid dynamics:
- Max Planck – University of Twente Center for Complex Fluid Dynamic;
- Eindhoven University of Technology;
- Physics of Fluids (PoF) group of Detlef Lohse, University of Twente (UT).

Together, Océ and the PoF-group revealed the disturbing role of bubbles entrained into piezo-acoustic ink channels and developed solutions for this problem.

Océ is a sponsor and research partner of the Fundamental Fluid Dynamics Challenges in Inkjet Printing (FIP) program

### DRIVEN BY FLUID DYNAMICS
Inkjet technology has a bright future. Océ printers are rapidly evolving, becoming ever faster and more versatile – largely thanks to a better understanding of how different and new types of fluid behave and can be manipulated. The science of fluid dynamics is one of the key drivers of innovation in the world of digital printing.

154

© Springer Nature Switzerland AG 2019
F. Alkemade, *A Century of Fluid Mechanics in The Netherlands*, https://doi.org/10.1007/978-3-030-03586-0_8

# 08

## EPILOGUE

What will fluid mechanics look like one hundred years from now, in 2118? History has taught us that such predictions are hard to make. Very few in this field around 1980 would probably have believed that some 25 years later impressive measurements of 3D flows would be possible, but developments in PIV and other techniques have indeed revolutionized the world of experimentation in fluid mechanics. Improvements in computer performance, in the refinement of CFD codes, and in the methods for visualizing digital data have been enormous in this period. On the other hand, in around 1980 quite a lot of researchers may have thought that 'the turbulence problem' would have been solved in the subsequent 25 years.

Sketching the future of fluid mechanics is tricky and usually amounts to an extrapolation of current research activities. Still, people in general (including scientists), do like to talk about the future from time to time. During the Burgersdag (Burgers day) in 2000, at that time the annual meeting organized by the Burgers Centre for all fluid mechanics physicists in the Netherlands, a panel discussion took place on several statements related to the future of turbulence research. The report of Nieuwstadt and Ooms on this meeting can be found in JMBC 15 years, published by the Burgers Centre in 2007.

One of the questions discussed in 2000, was: will experiments disappear? One of those attending remarked that there are indeed several less attractive aspects of real experiments. "Nevertheless, it has been strongly stressed that the role of the experiment either in the field or in the laboratory is far from being over. Apart from experiments in large facilities, e.g., to reach large Reynolds numbers, there are also many possibilities for small-scale experiments to illustrate or investigate certain aspects of turbulence dynamics. Historically, serendipity is more likely in the laboratory than in front of a screen. Therefore, the future for experimental work is bright, particularly if we take into account the instrumental techniques that are available to us. These allow us to go from the single-point data that has dominated turbulence research in the last century to multipoint data in the next century."

According to other attendees there was at that time "no single theoretical route towards a better understanding of turbulence and for the future this seems unchanged. ... The most important aim for theoretical work should be to provide concepts and ideas to guide the numerical and laboratory experiments." Another remark that was made: "As there is not a single theory of turbulence, one should not expect a universal turbulence model." The results obtained by researchers like Lohse and Westerweel and their co-workers since this panel discussion took place (see the remarks on Lohse's Physica lecture in § 5.1.2) support this supposition. It may be expected that in the coming years this view on turbulence will be generally accepted and will lead to new developments in turbulence research.

The Burgersdag was replaced a few years ago by the Burgers Symposium. During the 2018 edition the author put up two almost empty posters: one to ask attendees what they considered as landmark achievements in Dutch fluid mechanics in the past forty years (see chapter 1) and the other to ask them what (important) new developments in fluid mechanics they expect in the coming years (see below). As more or less a provocation the author wrote the first remark on this second poster: "Quantum computers make possible simulations of which we can only dream at the moment". The excellent CFD work which has been done in the Netherlands over the last decades gives confidence in a fruitful use of the possibilities of future computing tools like quantum computers (on which important research is performed in the Netherlands). The availability of already existing and of future experimental facilities, sometimes unique in the world, gives confidence that the future simulations can sufficiently be validated.

It is up to the author of the (completely digital?) volume on '150 years of fluid mechanics in The Netherlands' to evaluate which of the predictions of 2018 have come true.

# BURGERS SYMPOSIUM 2018

"BROAD APPLICATION OF FLUID MECHANICS IN MEDICINE AND HEALTH."

"ADVANCEMENTS IN BIOMIMICRY."

"NO DIFFERENCE BETWEEN EXPERIMENTAL AND NUMERICAL RESULTS (MEASURED INITIAL CONDITIONS FOR DNW)."

"SEAMLESS INTEGRATION OF 'PHYSICS AND CHEMISTRY' AND 'DATA SCIENCE' -> REALITY."

"MUCH MORE USE OF DNS (DIRECT NUMERICAL SIMULATIONS) AND LES (LARGE EDDY SIMULATIONS) FOR REALISTIC ENGINEERING FLOWS."

"ANALYTICAL SOLUTIONS OF THE NAVIER-STOKES EQUATIONS."

# GLOSSARY

**Aero-acoustics:** The study of sound generated by the interaction of flows with solid surfaces.

**AHD** = Laboratory for Aerodynamics and Hydrodynamics: The laboratory of the TH/TU Delft founded by Jan Burgers around 1920.

**Aquifer:** An underground layer of water-bearing permeable rock, rock fractures, or materials like gravel, sand, or silt.

**Boundary layer:** The thin region in the immediate vicinity of a boundary surface (such as a wall) where the viscous terms (e.g., diffusion) necessary to satisfy the no-slip condition come into play.

**BPM** = Bataafsche Petroleum Maatschappij: A company founded in 1907 which was a subsidiary of the Royal Dutch Shell company.

**Buoyancy:** The effects of differences in density in a fluid. One of these is Archimedes upward force on a body floating on or immersed in a liquid. Another is the convection flow in a gas which has a varying temperature.

**Cavitation:** Local 'cold boiling' of water at spots where the pressure drops to zero or negative values which leads to the formation of bubbles. The collapse of these bubbles causes noise, vibration, and damage to bodies in the vicinity.

**CFD** = Computational Fluid Dynamics: The application of numerical methods and computers in order to simulate and investigate fluid flows.

**Coherent structures:** In turbulent flows this is an organized component of the vorticity field which is phase-correlated over the entire space of this component. A well-known example is the hairpin vortex in the turbulent boundary layer.

**Colloids, colloidal suspensions:** Complex liquids which exist of mesoscopic particles (the colloids) in a solvent. The colloids are typically between 1 and 1000 nanometres, significantly larger than the size of the solvent molecules and small enough to show Brownian motion.

**Convection:** Convection is the coordinated, collective movement of groups of molecules within fluids, which leads to the transport of mass, momentum, and energy.

**Delta area:** This is the part of the Dutch province of Zeeland which is not bordering the North Sea. The area consists of islands, sea arms (e.g., the Oosterschelde), and lakes.

**Dispersion:** When particles move from places of high concentration to places of lower concentration, the term dispersion is used (which also has a lot of similar meanings in other fields of physics). Dispersion occurs regularly in fluids since usually the concentration in a fluid is not homogeneous.

**DNS** = Direct Numerical Simulations: A computer simulation in which the Navier-Stokes equations are numerically solved without the use of any turbulence model (as is used in LES).

**Drag:** A force acting opposite to the relative motion of any object moving with respect to a surrounding fluid (e.g., air resistance).

**Dredging:** The operation of removing material from one part of the water environment (e.g., the bottom of a river) and relocating it elsewhere.

**Fluidized bed:** A physical phenomenon occurring when a quantity of solid particles is placed under appropriate conditions to cause a solid/fluid mixture to behave as a fluid. This is usually achieved by the introduction of pressurized fluid through the particulate medium.

**Flume:** A human-made channel for water with walls raised above the surrounding terrain, e.g. used in hydraulic experiments.

**Flutter:** A dynamic instability of an elastic structure in a fluid flow, caused by positive feedback between the body's deflection and the force exerted by the flow.

**FOM** = Fundamenteel Onderzoek der Materie (Fundamental Research of Matter): A foundation, which was established in 1946 to stimulate and finance research related to physics in the Netherlands.

**Granular material:** A conglomeration of discrete solid, macroscopic particles (larger than 1 micrometre, usually in the order of 1 millimetre) characterized by a loss of energy whenever the particles interact.

**Groundwater flow:** The flow of water that is found underground in cracks and spaces in the soil, sand, and rocks.

**Hot-wire measurement technique:** The hot wire of an anemometer (velocity meter) is a heated by an electric current and cooled down by the colder fluid flow which passes by. The temperature of the hot wire is kept constant and the variations in the electric current are directly proportional to the flow velocity.

**Hydraulic head:** This is a concept that relates the energy in an incompressible fluid to the height of an equivalent static column of that fluid. In fact, it is a specific measurement for the pressure.

**Hydraulics:** In this volume this is the applied branch of fluid mechanics which is concerned with the free surface flow of water in 'natural' environments, usually on large scales (rivers etc.). Not to be confused with 'hydraulic machinery'.

**Hydrology:** The scientific study of the movement, distribution, and quality of water on Earth and other planets, including the water cycle and water resources.

**IUTAM** = International Union of Theoretical and Applied Mechanics.

**KNAW** = Koninklijke Nederlandse Akademie van Wetenschappen (Royal Netherlands Academy of Arts and Sciences).

**KNMI** = Koninklijk Nederlands Meteorologisch Instituut (Royal Netherlands Meteorological Institute).

**KSLA** = Koninklijke/Shell Laboratorium Amsterdam: The main (fundamental) research laboratory of Shell in the Netherlands.

**Laminar flow:** This type of flow occurs when a fluid flows in parallel layers, with no disruption between the layers. The flow can be described as smooth, non-chaotic (unlike turbulent flows).

**LDA** = Laser Doppler Anemometry: The technique of using the Doppler shift in a laser beam to measure the velocity in transparent or semi-transparent fluid flows.

**LES** = Large Eddy Simulation: A CFD method in which the smallest length scales of the flow, which are the most computationally expensive to resolve, are ignored via low-pass filtering of the Navier–Stokes equations. Such a low-pass filtering effectively removes small-scale information from the numerical solution but the effect of the small scales on the flow must somehow be modelled.

**Magnetohydrodynamics:** The study of the magnetic properties and the flow of electrically conducting fluids. Examples of such magneto-fluids include plasmas, liquid metals, salt water, and electrolytes.

**MARIN** = Maritiem Research Instituut Nederland (Maritime Research Institute Netherlands).

**MHD** = magnetohydrodynamics.

**Microfluidics:** A rather young field of fluid mechanics which deals with the behaviour, precise control and manipulation (e.g., mixing) of fluids that are geometrically constrained to a small, typically sub-millimetre, scale at which capillary penetration governs mass transport. Fluids in tiny structures behave differently from fluids in large volumes and flows are mostly laminar.

**Navier-Stokes equations:** Differential equations which describe every kind of motion of viscous fluids. These balance equations arise from applying Newton's second law to fluid motion, together with the assumption that the stress in the fluid is the sum of a diffusing viscous term and a pressure term.

**Newtonian fluids:** In these fluids the strain rate (or shear rate) does not have any influence on the viscosity; there is only one viscosity for the whole fluid (at a certain temperature). For these fluids the stress in the fluid is proportional to the 'deformation'. Examples of these fluids are 'simple' fluids like water and air.

**NIOZ** = (Koninklijk) Nederlands Instituut voor Onderzoek der Zee (Royal Netherlands Institute for Sea Research).

**NLL** = Nationaal Luchtvaart Laboratorium (National Aeronautics Laboratory).

**NLR** = Nationaal Lucht- en Ruimtevaart Laboratorium (National Aerospace Laboratory). Non-Newtonian fluids: see Newtonian fluids.

**NSMB** = Netherlands Ship Model Basin (see MARIN): English name for the NSP.

**NSP** = Nederlands Scheepsbouwkundig Proefstation (see NSMB): The initial name of MARIN.

**Offshore:** Usually refers to structures and facilities in a marine environment, e.g., oil platforms.

**Physical technology:** Physical treatments of materials by means of machines without changing the molecules, e.g., mixing.

**PIV** = Particle Image Velocimetry: An experimental method for visualizing and measuring flow fields. The fluid is seeded with small tracer particles. The fluid with entrained particles is illuminated so that particles are visible. The motion of the seeding particles is filmed, and the images can be used to calculate the velocity field (speed and direction) of the flow.

**Plasma:** The plasma state is different from the well-known ones (gas, liquid, solid). Like a gas, plasma does not have definite shape or volume. But plasmas are electrically conductive, produce magnetic fields, and electric currents, and respond strongly to electromagnetic forces.

**Polymer:** A large molecule, composed of many repeated subunits. They are found in both natural and synthetic materials (plastics).

**Potential flow:** An 'ideal' flow in which the velocity field is the gradient of a scalar function: the velocity potential. A potential flow is characterized by an irrotational velocity field, which is a valid approximation for several applications but cannot be used in other situations (e.g., turbulent flows).

**Reynolds number (Re):** Important dimensionless quantity in fluid mechanics used to help predict flow patterns in different fluid flow situations. It is the ratio of inertial forces to viscous forces. At low Re, flows tend to be laminar, while at high Re turbulence occurs.

**Rheology:** The study of fluid stress-strain relationships. Rheology generally accounts for the behaviour of non-Newtonian fluids, by characterizing the minimum number of functions that are needed to relate stresses with rate of change of strain or strain rates.

**Rijkswaterstaat:** The Directorate-General for Public Works and Water Management in the Netherlands. Among other things, it is responsible for the construction and maintenance of waterways.

**RSL** = RijksStudiedienst voor de Luchtvaart (Government Service for Aeronautical Studies): The initial name of the NLR.

**RWS** = Rijkswaterstaat

**Sediment:** Sediment is a naturally occurring material, usually granular like sand, that is broken down by processes of weathering and erosion. Sediments can be transported by water, air, or other fluids.

**Shock wave:** A type of propagating disturbance that moves faster than the local speed of sound in a medium (usually air). Like an ordinary wave, a shock wave carries energy and can propagate through a medium but is characterized by an abrupt, nearly discontinuous, change in pressure, temperature, and density of the medium.

**Slug flow:** A liquid–gas two-phase flow in which the gas phase exists as large bubbles separated by liquid 'slugs'.

**Soft matter:** Matter which has two main features: complexity and flexibility. These differentiate it from simple Newtonian fluids. Soft matter covers a wide range of materials, including colloids, foams, emulsions, polymers and gels, granular materials, liquid crystals and a certain number of biological materials.

**Stokes flow:** A type of fluid flow where advective inertial forces are small compared with viscous forces. The Reynolds number is very low. This is a typical situation in flows where the fluid velocities are very small, the viscosities are very large, or the length-scales of the flow are very small.

**Surging:** The transient sudden rise or fall of pressure in a pipeline.

**TH** = Technische Hogeschool (Technical University)

**TNO** = Nederlandse Organisatie voor Toegepast Natuurwetenschappelijk Onderzoek (Netherlands Organisation for Applied Scientific Research).

**Transition:** The process of a laminar flow becoming turbulent.

**TU** = Technical University. In the Netherlands there are two: Delft (TUD) and Eindhoven (TUE).

**Two-phase flow:** A flow in which two different states of matter are present, e.g., gas and liquid. These flows can occur in various forms, such as flows transitioning from pure liquid to vapor as a result of external heating, separated flows, and dispersed two-phase flows where one phase is present in the form of particles, droplets, or bubbles in a continuous carrier phase (i.e., gas or liquid).

**Turbulence model:** The equations governing turbulent flows can only be solved directly for simple cases of flow. For most real-life turbulent flows, CFD simulations use turbulent models to predict the evolution of turbulence. These turbulence models are simplified constitutive equations that predict the statistical evolution of the flows.

**Viscoelastic fluids:** Fluids that exhibit both viscous and elastic characteristics when undergoing deformation.

**Viscosity:** A fluid property that measures the fluid's resistance to shear stress. One could say that the viscosity is an indication of the 'thickness' or 'syrupiness' of a fluid.

**Vorticity:** A measure for the angular velocity (rate of rotation) in a fluid. A region in a fluid in which the flow revolves around an axis line (which may be straight or curved) is called a vortex.

**WL** = Waterloopkundig Laboratorium (later: Delft Hydraulics, today: Deltares).

# SOURCES

## MONOGRAPHS

Akker H van den, Kleijn Chr (eds) (1999) Het gewicht van de Witte Olifant – 50 jaar Kramers Laboratorium voor Fysische Technologie. s.n., s.l.

Allen J et al (1963) Selected aspects of hydraulic engineering. Technical University of Delft, Delft

Ankersmit W et al (eds) (2017) 175 jaar TU Delft. Histechnica, Delft

Bakker M, Hooff G van (1991) Gedenkboek Technische Universiteit Eindhoven 1956-1991. TU Eindhoven, Eindhoven

Bakker PG, Coene R, Van Ingen JL (1992) Essays on aerodynamics. Delft University Press, Delft

Baudet H (1992) De lange weg naar de Technische Universiteit Delft. Sdu, Den Haag

Beltman H (1975) Verdikken en geleren. Dissertation, Wageningen University

Berkel K van (1985) In het voetspoor van Stevin. Boom, Meppel

Biesheuvel A, Heijst GJF van (1998) In fascination of fluid dynamics. Kluwer Academic, Dordrecht

Bliek JA van der (1994) 75 Years of aerospace research in The Netherlands. NLR, Amsterdam

Boer JJ & Drukker JW (2011) High tech, human touch. Universiteit Twente, Enschede

Boersen SJ (2018) Albert Gillis von Baumhauer. Lanasta, Emmen

Bosch A, Ham W van der (1998) Twee eeuwen Rijkswaterstaat 1798-1998. Europese Bibliotheek, Zaltbommel

Braams CM et al (1971) Nederlands Tijdschrift voor Natuurkunde Jubileumnummer. Ned. Natuurkundige Vereniging, s.l.

Broeder JJ et al (1982) Physics in The Netherlands. FOM, Utrecht

Bunch B, Hellemans A (2004) The history of science and technology. Houghton Mifflin, Boston

Darrigol O (2005) Worlds of flow. Oxford University Press, Oxford

Dijkstra D et al (eds) (1998) Floating, flowing, flying. Kluwer Academic, Dordrecht

Dirkzwager JM (1977) Water. Martinus Nijhoff, Den Haag

Dommerholt A, Warmerdam P (2017) De geschiedenis van het hydraulica laboratorium. WUR, Wageningen

Drost S (2015) Extrusion instability in an aramid fibre spinning process. Dissertation, Delft University of Technology

Drummen M et al (eds) (2001) Evolutie in weer- en sterrenkunde - 100 jaar Nederlands onderzoek Stichting De Koepel, Utrecht

Eckert M (2006) The dawn of fluid dynamics. Wiley-VCH, Weinheim

Elsenaar A (2012) 50 Years high speed wind tunnel testing in The Netherlands. Foundation Historical Museum NLR, Amsterdam

Ende JCM van den (1994) The turn of the tide - computerization in Dutch society, 1900-1965. Delft University Press, Delft

Ferreiro LD (2007) Ships and science. MIT Press, Cambridge MA

Flitsen uit het K.N.M.I. (1979) Staatsuitgeverij, 's-Gravenhage

Forbes RJ, O'Beirne DR (1957) The technical development of the Royal Dutch/Shell, 1890-1940. Brill, Leiden

Götz HF, Tak CJM (1995) Gemalen. Tak Architektenbureau, Delft

Ham W van der (1999) Heersen en beheersen. Europese Bibliotheek, Zaltbommel

Ham W van der (2007) Verover mij dat land - Lely & de Zuiderzeewerken. Boom, Amsterdam

Havinga A et al (1959) Research inspired by the Dutch windmills. Veenman, Wageningen

Heiningen H van (1991) Diepers en delvers - geschiedenis van de zand- en grindbaggeraars. Walburg Pers, Zutphen

Heijn J et al (1984) Natuurkunde in Nederland. Ministerie van Onderwijs en Wetenschappen, s.l.

Henssen EWA (1995) Uit de geschiedenis der Nederlandse geologische wetenschappen. STYX, Groningen

Huijs F et al (2003) Het Scheepsbouwkundig Gezelschap "William Froude" vereeuwigd. S. G. "William Froude", Delft

International jubilee meeting on the occasion of the 40th anniversary of the Netherlands Ship Model Basin (1972). NSMB, Wageningen

Kasteel ThJ van et al (1957) Een kwarteeuw TNO 1932-1957. TNO, Den Haag

Korteweg J (2013) 70 jaar IHC Merwede. Boekschap, Arnhem

Lammeren WPA van (1952) The Netherlands Ship Model Basin. NSMB, Wageningen

Lintsen H, Schippers H (eds) (2006) Gedreven door nieuwsgierigheid. SHT / TUE, Eindhoven

Lumley JL et al (eds) (1996) Research trends in fluid dynamics. AIP Press, Woodbury

Malkezadeh R (2012) Severe slugging in gas-liquid two-phase pipe flow. Dissertation, Delft University of Technology

Meier GEA (ed.) (2000) Ludwig Prandtl, ein Führer in der Strömungslehre. Vieweg, Braunschweig/Wiesbaden

Natuurwetenschappelijk onderzoek in Nederland (1942). Noord-Hollandsche Uitgevers Maatschappij, Amsterdam

Nichol K (2011) Fluidization and fluctuations in granular systems. Dissertation, Leiden University

Nieuwstadt FTM, Steketee JA (eds) (1995) Selected papers of J.M. Burgers. Kluwer Academic, Dordrecht

Nieuwstadt FTM (s.a.) Stromingsleer in maatschappij en onderzoek. J.M. Burgerscentrum, Delft

Oost E van et al (eds) (1998) De opkomst van de informatietechnologie in Nederland. Ten Hagen Stam, Den Haag

Petersen AC (2012) Simulating nature – a philosophical study of computer-simulation uncertainties and their role in climate science and policy advice. CRC Press, Boca Raton

Prinsenmolencommissie (1942) Het Prinsenmolenboek. Veenman en Zonen, Wageningen

Reynhart AFA (1951) De proeffabrieken voor physische en chemische technologie van de Technische Hogeschool te Delft. Bataafsche Petroleum Maatschappij, s.l.

Rooijendijk C (2009) Waterwolven - een geschiedenis van stormvloeden, dijkenbouwers en droogmakers. Atlas, Amsterdam/Antwerpen

Rouse H, Ince S (1957) History of hydraulics. Iowa Institute of Hydraulic Research, s.l.

Schmid WLH (1930) Over de werking van de luchtlift voor water. Dissertation, TH Delft

Schot JW et al (ed) (1998) Techniek in Nederland in de twintigste eeuw I. Stichting Historie der Techniek, Eindhoven/Walburg Pers, Zutphen

Schot JW et al (eds) (2000) Techniek in Nederland in de twintigste eeuw II. Stichting Historie der Techniek, Eindhoven/Walburg Pers, Zutphen

Schweppe J (1989) Research aan het IJ. Shell Research, Amsterdam

Schootbrugge G van de (2013) Van PULSIM naar Fluid Dynamics. TNO, Delft.

Thijsse, JTh (1972) Een halve eeuw Zuiderzeewerken 1920-1970. Tjeenk Willink, Groningen

Thoenes D (1957) Stofoverdracht bij stroming door een vast bed van korrelig materiaal. Excelsior, 's-Gravenhage

Vandersmissen H (1998) Het woelige water - watermanagement in Nederland. Teleac/NOT, Hilversum / Inmerc, Wormer

Veen J van (1936) Onderzoekingen in de hoofden in verband met de gesteldheid der Nederlandse kust. Algemeene Landsdrukkerij, 's-Gravenhage

Veen J van (1962) Dredge – drain – reclaim. Martinus Nijhoff, The Hague

Verbong G, Berkers E, Taanman M (2005) Op weg naar de markt. ECN, Petten

Vries JJ de (1982) Anderhalve eeuw hydrologisch onderzoek in Nederland. Rodopi, Amsterdam

Wakker KF et al (eds) (2002) Delfts goud. Technische Universiteit Delft – Bèta Imaginations, Delft

Weber R (1998) The spirit of IJmuiden – fifty years of the IFRF. IFRF, IJmuiden

Wijnand JH (1938) Bouwende gravers. Amsterdamsche Ballast Maatschappij, s.l.

Wittenberg H et al (1990) Vijftig jaar 'Vliegtuigbouwkunde' in Delft. Delftse Universitaire Pers, Delft

Zhu Y (2006) Breach growth in clay-dikes. Dissertation, Delft University of Technology

## ARTICLES

Dop H van, Axelsen S (2007) Large eddy simulation of the stable boundary-layer: a retrospect to Nieuwstadt's early work. Flow Turbulence Combustion 79:235-249

Gerritsma J (1957) The Shipbuilding Laboratory of the Delft University of Technology. Int. Shipbuilding Progress 5:12-23

Maas HJ van der (1957) Aeronautical research in The Netherlands. J of the Royal Aeronautical Society 61: 295-334

Monna W, Bosveld F (2013) Het KNMI Cabauw observatorium bestaat 40 jaar. Meteorologica 22:4-8

Proceedings

Garbrecht G (ed) (1987) Hydraulics and hydraulic research. Balkema, Rotterdam & Boston

Hough JL (ed) (1951) Turbidity currents and the transportation of coarse sediments to deep water. Soc. of Economic Paleontologists and Mineralogists, Tulsa

Howden N, Mather J (eds) (2013) History of hydrogeology. CRC Press, Boca Raton

Kraus NC (ed) (1996) History and heritage of coastal engineering. ASCE, New York

Wesseling P (ed) (1987) Research in numerical fluid mechanics. 25th Meeting of the Dutch Association for Numerical Fluid Mechanics, Delft, 1986. Vieweg, Braunschweig

## REPORTS

Lammeren WPA van (1963) Facilities and experiment techniques at the Netherlands Ship Model Basin. N.S.M.B., Wageningen

Staatscommissie Lorentz (1926) Verslag van de Staatscommissie ... Zuiderzee. Algemeene Landsdrukkerij, 's-Gravenhage

Timmerhuis VCM (1999) Ruimte voor vernieuwing. AWT, Den Haag

Several inaugural and farewell lectures of professors of Dutch universities

Several necrologies and other articles which have appeared in Delta (TUD), Nederlands Tijdschrift voor Natuurkunde, and KNAW Yearbooks

Several annual reports of the J.M. Burgerscentrum

Several items on Wikipedia

Various documents from the J.M. Burgers Archives at the Section of Fluid Mechanics, Process & Energy, 3mE, TU Delft (Caroline Legierse)

## REFERENCES FOR THE CONTRIBUTION OF UNILEVER:

1 Dubbelboer A (2016) Towards optimization of emulsified consumer products: modelling and optimization of sensory and physicochemical aspects. Dissertation, Eindhoven University of Technology (available on-line via TU/e repository);

2 Haighton AJ (1976) Blending, chilling and tempering of margarines and shortenings. J Amer Oil Chem Soc 53:397-399;

3 Trommelen AM, Beek WJ (1971) Flow phenomena in a scraped-surface heat exchanger ("Votator type"). Chem Eng Sci 26:1933-1942;

4 Münüklü P (2005) Particle formation of ductile materials using the PGSS technology with supercritical carbon dioxide. Dissertation, Delft University of Technology (available on-line via TUD repository);

5 Janssen JJM, Hoogland H (2014) Modelling strategies for emulsification in industrial practice. Can J Chem Eng 92:198-202

## MUSEUMS VISITED

Haarlemmermeermuseum De Cruquius, Cruquius; Maritiem Museum, Rotterdam; Nationaal Baggermuseum, Sliedrecht; Scheepvaartmuseum, Amsterdam; Sonnenborgh, Utrecht; Nationaal Militair Museum, Soesterberg

## IMAGES

The sources of the images in this book are mentioned in the captions. If no source is mentioned, it concerns a photo which was taken by the author, sometimes of documents from his personal collection.

## THE SOURCES OF THE IMAGES AT THE OPENING OF THE MAIN CHAPTERS:

1 The photo of several documents consulted by the author was taken by him;

2 The aerial view of the Cruquius pumping station was kindly provided by Haarlemmermeermuseum De Cruquius (see www.haarlemmermeermuseum.nl for the new museum they are planning);

3 The photo of one of the ventilation buildings of the State Mines, around 1932 is in the Burgers Archives, courtesy of Delft University of Technology;

4 The photo of one of the facilities in the laboratory of Warmte en Stroming of the TH Eindhoven, taken around 1970, is from TU/e IN BEELD, courtesy of TU Eindhoven archives

5 A photo taken by Ilse Hoekstein-Philips during the Burgers Symposium in June 2018, organised by the J.M. Burgerscentrum (Leen van Wijngaarden is on the left);

6 Images from daily life at Burgers' laboratory in Delft around 1950 (with amongst others a very young Charles Hoogendoorn and Wim Welling, electronics expert for many experiments) from a photo album in the Burgers Archives, courtesy of Delft University of Technology.

7 Research at InnoSportLab De Tongelreep, a very special swimming pool in Eindhoven (photo by and courtesy of Jos Jansen Photography).

8 A prototype for a futuristic, partly Dutch and ill-fated spacecraft project. Photo taken by the author at the National Military Museum in Soesterberg.

The images accompanying the seven contributions of the sponsors of this volume (see Acknowledgements) are courtesy of the corresponding companies.

The cover photo (the pumping station Lely near Medemblik and its environment) was made by Siebe Swart / Hollandse Hoogte.

# ABOUT THE AUTHOR

Dr.ir. Fons Alkemade (1966) became involved in fluid mechanics as a student of Mechanical Engineering at the Delft University of Technology in 1987. After graduating he became a PhD student in 1989 at the Laboratory for Aerodynamics and Hydrodynamics, the lab which had been established by Jan Burgers around 1920. He witnessed the flowering of the 'stromingsleer' in The Netherlands and the birth of the Burgers Centre in the early 1990s. In 1994 his thesis On vortex atoms and vortons was defended before a committee chaired by professor Frans Nieuwstadt. During his PhD period he started to assess the archival material which had been deposited in Delft and in 1995 he published a biography of Burgers, for which he had interviewed a lot of people from the (former) fluid mechanics community. After 1994 Fons Alkemade became a freelance science writer, automotive historian, physics teacher and author/editor of schoolbooks on physics. Occasionally he kept doing (historical) research and writing on topics from fluid mechanics.